THE MOLECULAR GEOMETRIES OF COORDINATION COMPOUNDS IN THE VAPOUR PHASE

THE MOLECULAR GEOMETRIES OF COORDINATION COMPOUNDS IN THE VAPOUR PHASE

by

DR. MAGDOLNA HARGITTAI

and

DR. ISTVÁN HARGITTAI

Central Research Institute of Chemistry
of the Hungarian Academy of Sciences,
Budapest

ELSEVIER SCIENTIFIC PUBLISHING COMPANY
AMSTERDAM – OXFORD – NEW YORK 1977

Distribution of this book is being handled by the following publishers:
for the U. S. A. and Canada
Elsevier/North Holland, Inc.
52 Vanderbilt Avenue
New York, New York 10017

for the East European countries, China, Cuba, Korean People's Republic, Mongolia and People's Republic of Vietnam
Akadémiai Kiadó, Publishing House of the Hungarian Academy of Sciences, Budapest

for all remaining areas
Elsevier Scientific Publishing Company
335 Jan van Galenstraat
P. O. Box 211
Amsterdam, The Netherlands

Library of Congress Cataloging in Publication Data

Hargittai, Magdolna
  The molecular geometries of coordination compounds in the vapour phase.

  Translation of Koordinációs vegyületek gőzfázisú molekulageometriája.
  Bibliography: p.
  Includes index.
  1. Coordination compounds.   2.   Molecules.
I. Hargittai, István, joint author. II. Title.
QD474.H3713   541'.2242   76-16501

ISBN 0-444-99832-2
Copyright © *Akadémiai Kiadó, Budapest 1977*

Joint edition published by Elsevier Scientific Publishing Company, Amsterdam, The Netherlands, and Akadémiai Kiadó, The Publishing House of the Hungarian Academy of Sciences, Budapest, Hungary

Printed in Hungary

PREFACE

This book is the English version of a Hungarian book with the same title that appeared as Volume 23 in the series A Kémia Újabb Eredményei (Advances in Chemistry), edited by Béla Csákvári, Akadémiai Kiadó, Budapest, 1974. During its translation into English it was revised and augmented with some references to the literature that has been published meanwhile.

The book is intended for students and practising chemists who are interested in structural chemistry, particularly in the molecular structures of coordination compounds. We also hope that our data compilations and comments will serve as a useful addition to undergraduate courses in inorganic chemistry.

We are very indebted to Professor Csákvári who encouraged us at various stages in the preparation of this book.

Our gratitude goes to several colleagues and friends who read the manuscript or parts of it. Professor Victor P. Spiridonov reviewed several chapters and Dr. Denis A. Kohl read the entire manuscript. All of them made valuable comments and provided useful criticism.

Special thanks are due to Dr. Ann Schmiedekamp who acted as the language referee of the translation. Not only did she correct our errors in the English but she also participated creatively in shaping the text.

The painful job of typing the text was performed superbly by Mrs. Júlia Simon, and Mrs. Judit Szilágyi provided valuable help in drawing the figures.

Our appreciation is expressed also to Mrs. Krisztina Kállay for her careful editorial work and to Mr. Sándor Csonka for preparing the figures in their final form.

We are grateful for the kindness of all those workers in this field who sent us reprints and preprints of their papers. Most of all, we greatly appreciate their contributions to the subject matter of this book. We are especially grateful to Dr. Barbara Starck for many useful references.

Writing a book always places some burden on the family of the author. As some small compensation we would like to dedicate this book to our children, Balázs and Eszter.

Magdolna and István Hargittai

CONTENTS

INTRODUCTION

The scope of this book is restricted to the vapour-phase molecular geometries of coordination compounds. We feel that it is hardly necessary to stress the importance of information on molecular structures to other branches of chemistry. However, it is worthwhile to emphasize once again the special significance of results that originate from vapour-phase studies. Of particular importance is the information collected on coordination compounds, where the linkages between various parts of the molecule may be weak. The vapour phase is the only phase in which the intramolecular forces are exclusively responsible for the molecular structure. There are numerous instances where the structures of the vapour and condensed phases are considerably different.

The structure of a molecule consists of many features, of which only the geometrical aspects will be dealt with in detail. Obvious correlations with other characteristics, *e.g.*, electronic structure, will also be pointed out in some instances.

The selection of the classes of compounds that are described was somewhat arbitrary. The classification is explained in the next section. We have attempted to cover all of the electron diffraction and microwave spectroscopic structural determinations on these compounds up to the end of 1973, with some references from the beginning of 1974 also being included.

With improvements in experimental and computational techniques, the precision of the determination of geometrical parameters has increased considerably during recent years. An increasing amount of attention* is also being paid to the physical significance of the structural parameters determined by various techniques. These parameters differ because the intramolecular motion is averaged differently in each instance. We draw attention to such interpretational problems as we feel that they will become even more important in the future.

---

* Cf., *e.g.*, the Conference on Critical Evaluation of Chemical and Physical Structural Information, Dartmouth College, June 24–29, 1973. Abstracts. National Research Council, National Academy of Sciences, Washington, D. C., 1973.

GENERAL CONCEPTS

COORDINATION COMPOUNDS

A great variety of definitions for coordination compounds can be found in the literature (see, *e.g.*, Cotton and Wilkinson, 1972; Bersuker, 1971). According to one of the most general definitions, the molecule of a coordination compound consists of a central atom and its surrounding ligands. A more restricted version defines coordination compounds as those whose central atom and ligands may exist separately in chemically non-extreme conditions. The former definition is applicable to almost all compounds, while the latter excludes some that are usually considered to be typical coordination compounds.

Further modifications of the definitions are used. One is that the coordination molecule consists of a central atom and ligands in which the central atom has not yet attained its maximum possible coordination number. Another is that the central atom is linked to more ligands than would correspond to the classical concept of valency. Molecules that are formed from two or more smaller molecules are also considered to be coordination compounds. This definition might seem to obscure the boundary between coordination compounds and polymers in some instances, but this problem is not relevant in the present context as only vapour-phase studies are to be considered.

Because of the particular subject matter of this book, the lack of a strict definition does not cause any difficulties. The number of vapour-phase determinations of molecular geometry that have been made is not yet too large, although it is increasing very rapidly. Future compilations with similar scope may find it necessary to use stricter criteria. However, it was felt that at present the concept of coordination compounds could be regarded liberally. Accordingly, we included all classes of compounds that were referred to as coordination compounds at various places in the literature.

It may be interesting to note that it was vapour-phase studies that established the coordination nature of some compounds that previously were not considered to belong to this class.

We are fully aware of the possibility of strong disagreement about the selection of the classes of compounds treated, particularly in the borderline cases. Classification of the available compounds into the selected classes seemed to be a less ambiguous task, greatly facilitated by the bonding properties of the complexes. It is our hope that apart from any differences in opinion concerning selection and classification, the general picture given here is a correct and satisfactory representation of the present status and results of vapour-phase determinations of the molecular geometries of coordination compounds.

THE IMPORTANCE OF VAPOUR-PHASE DATA

As already mentioned briefly, the special importance of vapour-phase data on molecular structure arises from the fact that they refer to the free, isolated molecule. As the molecules are far apart in the vapour phase, their interaction can be ignored. This is not so, however, in condensed phases, including solutions, where most of the studies in coordination chemistry have been and are expected to be performed. Packing considerations in the crystal phase and solvent effects in solutions, to mention only two of the important factors, may result in the formation of structures different from those which would be present if only intramolecular forces were acting. We will mention only a few of the different structures that have been established for the different phases.

The importance of intermolecular interactions in the condensed phases increases when weaker interactions can play a significant role in shaping the molecular geometry. It is well known that a change in the length of a bond usually requires more energy than an angular deformation. The largest deviations in the condensed phase molecular geometries, compared with the vapour-phase structures, may be expected for molecules with a low barrier to internal rotation, intramolecular hydrogen bridges or coordination-type linkages. As the theoretical considerations on the bonding in coordination compounds have been developed for isolated molecules (cf., Mulliken and Person, 1969), the vapour phase is the most appropriate medium for comparison (Downs, 1968).

Many types of differences have been recorded between the molecular structures referred to the vapour and condensed phases. For some salts, such as, for example, sodium chloride, an entirely different type of structure is formed in the crystal (a face-centred cubic ionic lattice) from that in the vapour phase (NaCl, Na_2Cl_2 molecules). In other instances the molecular configuration is changed. As an example, we mention the structures of B_2Cl_4 in the vapour (D_{2d} symmetry) and crystal (D_{2h} symmetry) phases:

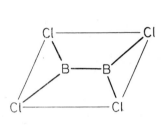 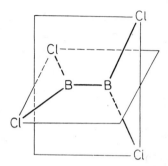

As can be seen, the first configuration differs from the second by the rotation of the BCl_2 group around the $B-B$ bond. For the sake of completeness, the symmetries of the B_2X_4 molecules, as determined in three phases, are presented below:

	Vapour	Liquid	Crystal
B_2F_4	D_{2h} (a)	D_{2d} (b)	D_{2h} (c)
B_2Cl_4	D_{2d} (d)	D_{2d} (e)	D_{2h} (f)
B_2Br_4	D_{2d} (g)	D_{2d} (h)	D_{2d} (h)

(a) Vibrational spectroscopy, Durig, Thompson, Witt and Odom (1973); Electron diffraction, Hedberg (1974). Earlier studies concluded D_{2d} symmetry for B_2F_4 (vibrational spectroscopy, Finch, Hyams and Steele (1965)).

(b) Vibrational spectroscopy, Gayles and Self (1964); Nimon, Seshadri, Taylor and White (1970).

(c) X-ray diffraction, Trefonas and Lipscomb (1958); Vibrational spectroscopy, Durig, Thompson, Witt and Odom (1973).

(d) Electron diffraction, Ryan and Hedberg (1969).

(e) Vibrational spectroscopy, Durig, Saunders and Odom (1971).

(f) X-ray diffraction, Atoji, Wheatley and Lipscomb (1955; 1957).

(g) Extrapolated from the other data.

(h) Vibrational spectroscopy, Odom, Saunders and Durig (1972).

13

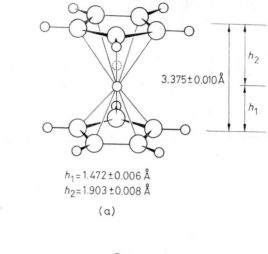

$h_1 = 1.472 \pm 0.006$ Å
$h_2 = 1.903 \pm 0.008$ Å

(a)

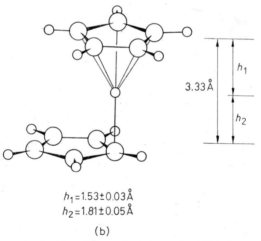

$h_1 = 1.53 \pm 0.03$ Å
$h_2 = 1.81 \pm 0.05$ Å

(b)

Fig. 1. The molecular configuration of bis(cyclopentadienyl)beryllium (a) in the vapour phase (Almenningen, Bastiansen and Haaland, 1964; Haaland, 1968); (b) in the crystalline phase (Wong, Lee, Chao and Lee, 1972)

The bond lengths and bond angles are known in both the vapour and crystal phases for B_2Cl_4 only and the differences are not significant, although the uncertainties are relatively large as seen from the following data.

	$r(B-Cl)$, Å	$r(B-B)$, Å	$< Cl-B-Cl$
Vapour $(d)^*$	1.750 ± 0.011	1.702 ± 0.069	$118.65 \pm 0.66°$
Crystal $(f)^*$	1.73 ± 0.02	1.75 ± 0.05	$120.5 \pm 1.3°$

* For references see p. 13.

The change in the molecular geometry of bis(cyclopentadienyl)-beryllium is particularly fascinating as the vapour-phase and crystal-phase data are compared. According to the electron diffraction results of Almenningen, Bastiansen and Haaland (1964) and Haaland (1968), the two cyclopentadienyl rings in gaseous $(C_5H_5)_2Be$ are parallel and staggered, and the distance between them is 3.375 ± 0.010 Å. The beryllium atom is situated on the coinciding five-fold symmetry axes of the two C_5H_5 rings and may occupy two alternative positions, at $h_1 = 1.472 \pm 0.006$ Å from one ring and $h_2 = 1.903 \pm 0.008$ Å from the other. The $(C_5H_5)_2Be$ molecule has C_{5v} symmetry and the structure is shown in Fig. 1. The molecular configuration in the crystal phase differs considerably from that in the vapour phase. An X-ray diffraction study by Wong, Lee, Chao and Lee (1972) showed that the distance between the two parallel cyclopentadienyl rings is 3.33 Å. One of the rings is shifted, however, from the symmetry axis on which the Be atom is located. As shown in Fig. 1, the distance of the beryllium atom from one of the rings, under which it lies, is 1.53 ± 0.03 Å and the distance between the Be atom and the plane of the other ring is 1.81 ± 0.05 Å. The structure elucidated for the crystal phase was not found to be compatible with the gaseous diffraction data (Wong, Lee, Chao and Lee, 1972; Drew and Haaland, 1972a).

The beryllium atom oscillates between the two alternative positions along the five-fold rotation axis in the vapour phase. In the crystal, this motion slows as the temperature is decreased (the crystal structure study was carried out at $-120°C$). The linkage is very weak between the beryllium atom and the more distant ring. Thus, the shift of this C_5H_5 ring may be explained by the influence of intermolecular forces.

DETERMINATION OF MOLECULAR GEOMETRY BY PHYSICAL TECHNIQUES

The geometry of a molecule can be characterized in a qualitative way by the shape and symmetry of the molecule. A more quantitative characterization consists of the relative three-dimensional positions

15

of the atoms that constitute the molecule or, more descriptively, the bond distances, bond angles and angles of internal rotation. These two types of characterization are often called the two levels of molecular geometry and it is generally important to determine the molecular geometry on both levels. Our detailed discussion, however, will be concentrated mainly on the second level with supplementary notes on the first. The second level of molecular geometry is found to be particularly challenging, as observations on the variations in bond distances and bond angles for a series of compounds are believed to be one of the most effective methods of investigating the nature of chemical bonding.

The changes in bond distances and bond angles in a series of compounds are seldom very large. The interpretation of subtle variations in the geometrical parameters requires careful consideration of the physical significance of the parameters in question. Before summarizing the representations of molecular geometry, the physical techniques used for the determination of molecular structure will be surveyed.

The following compilation of physical techniques gives information concerning their capabilities for determining only the first or both levels of molecular geometry:

Technique	Molecular shape and symmetry	Geometrical parameters
Microwave spectroscopy	+	+
Rotational Raman spectroscopy	+	+
Vibrational Raman and infrared spectroscopy	+	−
Electron diffraction	+	+
X-ray diffraction	+	+
Neutron diffraction	+	+
Measurement of dipole moment	+	−
Magnetic measurements	+	−
Nuclear magnetic resonance spectroscopy	+	−
Mass spectrometry	+	−

This book does not aim to present these physical techniques and their applications in detail. An excellent introductory survey is provided by the book by Wheatley (1959). Some references to more advanced texts will also be mentioned.

Only the two main techniques for determining the vapour-phase molecular geometry, *viz.*, electron diffraction and microwave spectroscopy, are described to some extent in subsequent sections of this book. X-ray diffraction, the most widely applied technique for investigating crystal-phase molecular geometry, together with neutron diffraction, will be mentioned only in connection with the physical significance of the geometrical parameters that they provide. Of the other methods, one specific feature from each of the topics, vibrational spectroscopy and dipole moment measurements, will be considered.

With respect to the comparisons made between electron diffraction vapour-phase data and X-ray diffraction crystal-phase data, one must pay attention to more than the difference between the vapour and crystal phases. An important source of deviations between the geometrical parameters determined by the two techniques is the difference in the respective scattering interactions. It is sometimes observed that the structural parameters obtained by X-ray and neutron diffraction for the same crystalline compound differ by more than the limits of experimental error. As X-rays are almost entirely scattered by the electrons of the atom, X-ray diffraction provides the electron density distribution. Accordingly, the interatomic distances determined correspond to the distances between the centres of gravity of the electron density distribution. Neutron diffraction, on the other hand, provides the density distribution of the atomic nuclei. The third diffraction technique, electron diffraction, is capable of determining the charge density distributions of atoms and molecules and yielding information primarily on the positions of the atomic nuclei.

A practical example is thought to be the best means of illustrating the differences. The results obtained with the early techniques of electron diffraction (Hedberg, Jones and Schomaker, 1951, 1952; Jones, Hedberg and Schomaker, 1953) and X-ray diffraction (Nordman and Lipscomb, 1953a, 1953b) for the length of the $B-H$ bond in diborane, B_2H_6, differed considerably. The values for both terminal and bridge bond distances, $r(B-H)$, were found to be 0.1 Å less by X-ray diffraction than by electron diffraction. In up-to-date X-ray diffraction studies (Adrian and Feil, 1969), it is also usual to find shorter $B-H$, $C-H$ or $N-H$ bond lengths than are found by other methods. The following bond distances and bond angles resulted from the modern X-ray diffraction study on diborane by Smith and Lipscomb (1965):

$$r(B-B) = 1.76_2 \pm 0.01 \text{ Å}$$

$$r(B-H_b) = \begin{cases} 1.24 \\ 1.25 \end{cases} \pm 0.02 \text{ Å} \qquad r(B-H_t) = \begin{cases} 1.09 \\ 1.06 \end{cases} \pm 0.02 \text{ Å}$$

$$<H_t-B-H_t = 121.6 \pm 1°,$$

where H_b and H_t are the bridging and terminal hydrogen atoms, respectively. For comparison, the reader is referred to the data in Table 14, obtained from an up-to-date electron diffraction study. It was pointed out by Jones and Lipscomb (1969) that the origin of the discrepancy is caused by the failure of the spherical atom model used in the interpretation of the X-ray diffraction data. The electron density distribution corresponding to the bonding of the atomic pair $B-H$ causes an asymmetry of the electron density around the atoms, particularly around the hydrogen atom, and the resulting distance parameter is smaller than that which would correspond to the positions of the atomic nuclei.

ELECTRON DIFFRACTION*

This method is based on the phenomenon that a beam of fast electrons is scattered by the potential resulting from the charge distribution in the molecule. The resulting interference pattern depends on the molecular geometry. The structure-dependent part of the scattered electron intensity for a molecule with N atoms is given by the equation

$$M(s) = \text{const.} \sum_{\substack{i=1 \\ i \neq j}}^{N} \sum_{j=1}^{N} g_{ij}(s) \exp\left(-\frac{1}{2} l_{ij}^2 s^2\right) \sin\left[s(r_{ij} - \kappa_{ij} s^2)\right]/r_{ij}.$$

The internuclear distances are denoted by r_{ij}, their associated mean amplitudes of vibration by l_{ij} and the asymmetry parameters by κ_{ij}. The functions $g_{ij}(s)$, containing the electron-scattering amplitudes, can

---

* The following recent review articles are brought to the attention of the reader: Bartell (1972), Bauer (1970), Hargittai (1974), Karle (1973), Kuchitsu (1972) and Seip (1973).

18

be calculated theoretically and are assumed to be known in the usual structural determinations. Then, the electron diffraction structural analysis can be similarly characterized as the determination of the frequencies and the damping of the components of a sum of sine functions. The asymmetry parameter (κ_{ij}) in the frequency-modulating factor is related to the anharmonicity of the molecular vibrations. The angular variable s is given by $s = 4\pi\lambda^{-1} \sin \vartheta/2$, where λ is the electron wavelength and ϑ is the scattering angle.

The diffraction pattern, usually recorded experimentally on photographic plates, is produced when an electron beam crosses a beam of free molecules. In early days the positions and relative intensities of the maxima and minima of the interference pattern were estimated visually ("visual technique"). In modern experiments, a rotating sector is introduced into the path of the scattered electrons in order to compensate for the steeply falling background, otherwise the intensity would exceed the exposure limits of the photographic plate. The optical density distribution of the sectored diffraction pattern can then be accurately determined with a microphotometer. This experimental development was termed the sector-microphotometer technique. Today, this term has a wider meaning and usually signifies all the experimental, computational and theoretical developments of the electron diffraction technique.

Electron diffraction structural analysis is usually started by establishing the overall configuration of the molecule if it is not known from other sources. In doing this, the Fourier transform of the molecular intensity:

$$f(r) = \text{const.} \int_0^{s_{\max.}} sM(s) \exp(-as^2) \sin(sr)\, ds$$

is usually calculated. This function, called the radial distribution function, $f(r)$, consists essentially of Gaussian peaks corresponding to the individual internuclear distances. The radial distribution function is simply related to the probability distribution function, $f(r) = \Sigma P_{ij}(r)/r$, where $P_{ij}(r)\, dr$ is the probability that the distance between the ith and jth atoms has a value in the interval r and $r + dr$. The value of $s_{\max.}$ is the upper limit of the observed intensities and the value of a in the damping factor, $\exp(-as^2)$, is chosen in order to make the integral converge at $s_{\max.}$.

2*

The refinement of the structural parameters is based on comparisons of the experimental molecular intensities and radial distributions with the corresponding theoretical functions calculated for various models. The least-squares technique is employed in most modern studies for determining molecular intensities, yielding the standard deviations and correlation coefficients, in addition to the "best" set of parameter values.

The values that result directly from the electron diffraction structural analysis are obtained in terms of r_a parameters. The r_a internuclear distance is an effective parameter rigorously equal to the position of the centre of gravity of the $P(r)/r$ distribution, and is also called $r_g(1)$.* The position of the maximum of the $P(r)/r$ distribution is called r_m.

MICROWAVE SPECTROSCOPY**

Using this technique, the molecular geometry is determined from the pure rotational spectrum, which is obtained from dipole transitions between quantized rotational states of the molecule without a change in its vibrational state. The rotational constants are obtained from the frequencies of the rotational transitions, which are in the microwave region. As the rotational constants are related to the principal molecular moments of inertia, the molecular geometry can be deduced. For the simplest molecules, it is also possible to obtain the equilibrium values of the rotational constants from which the equilibrium molecular geometry (r_e structure) can be obtained. In a more general case, the r_0 internuclear distances, effective parameters with no well defined physical significance can be obtained. The r_0 parameters are usually derived from a least-squares fit of the bond distances and bond angles, or rather from a fit of the moments of inertia determined by them, to the observed quantities ($I = h/8\pi^2 B$, where I is the moment of inertia and B the rotational constant).

---

* The notations r_a and $r_g(1)$ are usually applied if the parameter originates from refinements based on the molecular intensities and the radial distribution, respectively.

** See the following monographs and review articles: Wilson and Lide (1955), Sugden and Kenney (1965), Wollrab (1967), Lide (1969), Gordy and Cook (1970).

Up to three rotational constants can be obtained from a given isotopic species, yielding three independent measurements. The molecular geometry of molecules larger than the simplest is determined, however, by a larger number of independent parameters. Accordingly, structural information provided by other techniques or further data from the rotational spectra of isotopically substituted species have to be utilized. The latter is possible because the equilibrium internuclear distances are unchanged by the isotopic substitution, while there is usually an appreciable change in the atomic masses, *i.e.*, in the moments of inertia. The procedure is complicated by the fact that the structure in the ground vibrational state is different from the equilibrium structure because of the zero-point vibrations. As the molecular vibrations are also mass dependent, the influence of the zero-point vibrations will be different for various isotopically substituted species. Therefore, r_0 parameters determined in this way will not be invariant to the particular choice of isotopic substitutions performed. The difficulties can be demonstrated well by instances where it is possible to investigate more than the minimum number of isotopically substituted species. The values of the r_0 parameters may differ even by 0.02 Å from each other or from parameters with a well defined physical meaning such as the equilibrium distance parameters. The problem is particularly severe for molecules with considerable intramolecular motion.

The substitution (r_s) structure, although it has no well defined physical meaning, is a very good approximation to the equilibrium structure. The following procedure leads to the determination of the r_s structure. First, the rotational constants, A_0, B_0 and C_0, are determined for the parent molecule, then one of the atoms is isotopically substituted and the rotational constants, A_0', B_0' and C_0', of the new molecule are determined. The coordinates of the substituted atom can then be calculated, with respect to the principal axis system of the parent molecule, from the differences $A_0' - A_0$, $B_0' - B_0$ and $C_0' - C_0$. The coordinates obtained in such a way are called substitution coordinates. As the procedure is repeated for every atom in the molecule, the substitution coordinates for all atoms are obtained and the r_s substitution structure emerges. If the substitution coordinates are determined for all but one atom, with the supplementary use of the centre-of-mass equations, $\Sigma m_i a_i = 0$, a structure essentially the same as the r_s structure can be obtained. If the substitution coordinates are missing for two atoms, the additional use of the $\Sigma \, m_i(a_i^2 + b_i^2) = I_c$ relationships make the determination of the geometry possible. However, the structure no longer has the meaning of the r_s parameters; it is an r_0 structure.

LIMITATIONS AND UNCERTAINTIES

We will briefly mention only some of the most important limitations of the applications of electron diffraction and microwave spectroscopy. The most crucial and frequent sources of error will also be listed in order to facilitate judgement of the reliability of the structural parameters. It should be stressed, however, that more complete treatises on these two techniques (see the references given previously) must be consulted for a more thorough evaluation of original reports on structural determinations.

A prerequisite for the application of both techniques is, of course, that it is possible to obtain an adequate vapour pressure of the compound to be investigated. The pressures required in the electron scattering experiments are about 10–20 torr, but pressures three orders of magnitude less are required in the microwave spectroscopic experiments. The difficulties encountered in studying compounds with low volatility will be indicated separately. For structural analysis by either technique, it is advantageous to collect the experimental data at the lowest possible temperature.

For microwave spectroscopy, it is not advantageous to study molecules with heavy atoms that undergo low-frequency motion as the population in the ground state is decreased. Another difficulty arises when studying molecules with atoms that have a nuclear quadrupole moment because splitting of the rotational transitions arises.

Microwave spectroscopy is limited to polar molecules. No analogous limitation exists for electron diffraction; on the contrary, the structural elucidation of molecules with high symmetry is greatly facilitated. An important limitation to both techniques is that the electron diffraction radial distribution and the rotational spectrum become increasingly complex for larger molecules and it is no longer possible to make reliable assignments. The possibility of reproducing the experimental data equally well with various models is much greater in the electron diffraction studies than in the microwave spectroscopic determinations. While it is true that microwave spectroscopy is capable

22

of handling only simpler molecules, the molecular configuration can be established with a high degree of certainty. The pattern of the rotational spectrum is a very sensitive function of molecular geometry and atomic mass; changes in the structural parameters or changes in the order of as few as two atoms may cause an entirely different spectrum to appear.

Closely packed internuclear distances in a molecule create strong correlations among the parameters for the electron diffraction analysis but have no effect on the microwave spectroscopic study. On the other hand, it is difficult to determine the position of an atom by microwave spectroscopy if it lies near one of the principal axes. Indeed, because of the various limitations, each technique may be able to provide only partial information on the molecular geometry. The effects of some of the limitations can be reduced by a combination of diffraction and spectroscopic measurements.

It is relatively easy to estimate the experimental error for either technique. On the other hand, it is very difficult to give a realistic measure of the total uncertainty of the structural determination. The quantities that are directly measured in the experiments are far removed from the structural parameters and the estimation of the errors and their propagation throughout the structural analysis is very complex. As modern experimental, computational and theoretical approaches become increasingly sophisticated, more and more sources of error become apparent and have to be taken into consideration. Hence it appears to be extremely important to list all the errors that have been taken into account and to examine carefully the process for the estimation of the uncertainties as structural data originating from different times and different workers are compared.

The experimental error in the electron scattering experiment is determined primarily by the errors in the values of the electron wavelength and the distance between the nozzle and the photographic plate. This error influences the uncertainty in determining the molecular size, i.e., the internuclear distances. As usual, the experimental error cannot be neglected when compiling the total error of a determination, unless it is corrected according to suitable spectroscopic standards. It is the determination of the mean amplitudes of vibration that is most strongly influenced by errors in the scattering functions applied and by the errors connected with various stages of the photographic process.

Extremely high resolutions can be obtained in microwave spectroscopy. Accordingly, high accuracy can be achieved in measuring the frequencies of the rotational transitions and calculating the rotational constants and moments of inertia. This accuracy is usually not the limiting factor in the determination of the geometrical parameters, with the exception of the simplest molecules. It is more important that the molecules are not rigid bodies and appreciable errors are introduced into the parameters by the presence of intramolecular motion and by the inadequacy of the corrections made for it.

Important, but particularly difficult for the reader of structural papers to perceive, are the errors caused by various assumptions and constraints used in the structural analyses for either technique. A critical approach is again called for in this regard.

Even in such a brief and superficial survey as is given here, we feel that it is important to indicate the role of impurities in the structural determinations. For electron diffraction, it can be stated, in summary, that the interfering effects of the impurities increase as the scattering power, *i.e.*, the atomic number, increases, and also as the internuclear distances in the impurity molecules approach those of the molecule to be studied. On the other hand, if the internuclear distances belonging to the two different molecules (the one to be studied and the other considered to be the impurity) are well separated in the radial distribution, the simultaneous determination of both structures may be possible. The presence of impurity molecules with internuclear distances similar to those of the compound being studied may considerably reduce the precision of the electron diffraction determination. The accuracy of the determination of the structure from the rotational spectrum is influenced by impurities only if they prevent the assignment of the spectrum. For example, impurities that consist of nonpolar molecules have no effect on the rotational spectrum. On the other hand, impurity molecules that have a rich spectrum in the region of interest may complicate the spectrum to such an extent that the assignment is impossible. Of the two methods, the microwave spectroscopic results are much less subject to misinterpretation due to the presence of impurities.

Numerous simple inorganic compounds have a rich variety of molecular species in their vapours, and many of these species can be considered to be coordination molecules. The studies of Brewer, usually considered to be the pioneering work in this field, date back to the discovery of the presence of Cu_3Cl_3 molecules in the vapour of copper(I) chloride (Brewer and Lofgren, 1950).

The high-temperature conditions are the common distinguishing feature of any experimental technique applied to studies of the molecular structures of compounds of low volatility. There is no definition, of course, of the range called "high temperature". In structural studies, however, temperatures above 300–500°C are definitely called high temperatures. High-temperature vapours can also be defined as vapours of compounds "that are in a solid phase at ordinary temperatures and pressures" (Akishin, Rambidi and Spiridonov, 1967).

The importance of mass spectrometry continues to increase in high-temperature studies of inorganic compounds (Grimley, 1967; Margrave, 1968), in the identification of molecular species, in the determination of the stability and the connecting order of atoms, and also in testing hypotheses concerning the shapes of the molecules. It is expected (Shol'ts and Sidorov, 1972) that this technique will eventually play a role in inorganic chemistry similar to that which it already plays in the study of organic compounds.

Up to the present time, microwave spectroscopy at high temperatures has been applied to studies of diatomic and some of the simplest polyatomic molecules (Kuczkowski, Lide and Krisher, 1966; Lide and Kuczkowski, 1967; Tiemann, Hoeft and Törring, 1972). Some special problems arise when this method is used. One of the requirements is the construction of an absorption cell that is able to operate at elevated temperatures. The interpretation of the spectra of high-temperature species of polyatomic molecules is more difficult because of the large number of molecules in excited vibrational states whose rotational spectra are also present. The intensity of the transition referred to the ground vibrational state is also decreased under these conditions. On the other hand, the structural data obtained for several vibrational states provide the possibility of extrapolating to the equilibrium geometry (r_e distances), at least for some simple molecules. It is relevant here to emphasize an important difference between electron diffraction and spectroscopy. The electron scattering intensities are obtained as averages from the molecules distributed among the vibrational states, while the rotational transitions belonging to various vibrational states appear separately.

Most of the high-temperature electron diffraction studies have been performed at the Moscow State University (Frost, Akishin, Gurvich, Kurkchi and Konstantinov, 1953; Akishin, Vinogradov, Danilov, Levkin, Martinson, Rambidi and Spiridonov, 1958; Akishin, Rambidi and Spiridonov, 1967; Vilkov, Rambidi and Spiridonov, 1967; Szpiridonov, 1972), the High Temperature Institute of the Academy of Sciences of the U.S.S.R. and some other laboratories (Bauer, Ino and Porter, 1960; Morino, Ukaji and Ito, 1966a, 1966b; Hedberg, 1973; Tremmel, Ivanov, Schultz, Hargittai, Cyvin and Eriksson, 1973). Here again, in addition to the special requirements of the instrumentation (the most important requirement being the high-temperature nozzle system), interpretational difficulties have to be overcome. Owing to the large amplitudes of vibration, the contribution of some of the distances to the molecular intensity, especially those which involve light atoms, is rapidly damped and consequently important features of the intensity distribution vanish at small values of the scattering angles.* Because of these effects and further interpretational difficulties, the structural parameters that originate from high-temperature electron diffraction experiments are usually less precise than those obtained from ordinary studies.

The determination of the molecular geometry of high-temperature species, in particular, requires the supplementary application of different techniques. The importance of two of those which serve to elucidate the molecular shape and symmetry is stressed here. One of these techniques is the determination of the permanent dipole moment by *electric deflection of molecular beams* with mass spectrometric detection (Wharton, Berg and Klemperer, 1963; Büchler, Stauffer and Klemperer, 1964a, 1964b). Important information concerning the molecular configuration is obtained, as only polar molecules are deflected by an inhomogeneous electric field. It is advantageous to use a mass spectrometer as the detector because the constitution of the high-temperature vapour is not obvious. Some of the results obtained using this technique are presented in Table 1.

*This is usually called the "washing out effect" in the Russian language literature (Rambidi and Spiridonov, 1964).

TABLE 1

*Results of studies on the electric deflection
of molecular beams*

Polar		Non-polar		References
LiF		Li_2F_2	(planar)	(a)
LiCl		Li_2Cl_2	(planar)	(a)
LiBr		Li_2Br_2	(planar)	(a)
LiI		Li_2I_2	(planar)	(a)
LiO		Li_2O	(linear)	(a)
		BeF_2	(linear)	(a)
		$BeCl_2$	(linear)	(a)
MgF		MgF_2	(linear)	(a)
CaF_2	(bent)			(b)
BaF_2	(bent)			(c)
BaI				(c)
BaI_2	(bent)			(c)
		ZnF_2	(linear)	(a)
		$ZnCl_2$	(linear)	(a)
		CdF_2	(linear)	(a)
		HgF_2	(linear)	(a)
		$HgCl_2$	(linear)	(a)
		HgI_2	(linear)	(a)
$PbCl_2$	(bent)			(a)
PbI				(a)
PbI_2	(bent)			(a)
		MnF_2	(linear)	(b)
		$MnCl_2$	(linear)	(b)
		CoF_2	(linear)	(b)
		NiF_2	(linear)	(b)
SmF_2	(bent)			(d)
EuF_2	(bent)			(d)
YbF_2	(bent)			(d)

(a) Büchler, Stauffer and Klemperer (1964a).
(b) Büchler, Stauffer and Klemperer (1964b).
(c) Wharton, Berg and Klemperer (1963).
(d) Kaiser, Falconer and Klemperer (1972).

The application of the *matrix isolation technique* in spectroscopy (Becker and Pimentel, 1956) also has special importance for the structural studies of high-temperature species (Linevsky, 1961). This technique is described in a recent book edited by Hallam (1973), which also includes the significant advances in studies of high-temperature species by infrared spectroscopy (Snelson, 1973) and Raman

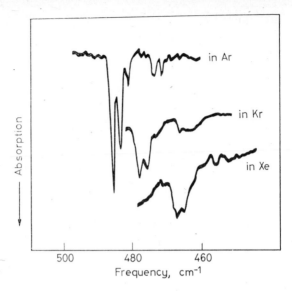

Fig. 2. Portions of the infrared spectra of cobalt(II) chloride recorded in various inert gas matrixes (Thompson and Carlson, 1968)

spectroscopy (Ozin, 1973). In these experiments, the high-temperature vapour is introduced into the matrix by simultaneous condensation of the matrix gas and the molecular beam of the species to be investigated. Special attention is required in order to examine the possibilities of interactions between the matrix and the guest molecules. The spectrum may change as various matrix gases are used (see, *e.g.*, Eliezer and Reger, 1972/73). A portion of the infrared spectra of cobalt(II) chloride recorded in argon, krypton and xenon is reproduced in Fig. 2 from the study of Thompson and Carlson (1968). The frequency shift is roughly proportional to the square of the atomic mass of the inert gas. The frequency of the argon matrix is the closest to the gas-phase value.

A few examples illustrate recent results in establishing the molecular configuration of coordination compounds by matrix isolation vibrational spectroscopy. According to an infrared study by Thompson and Carlson (1968), in addition to the linear $(D_{\infty h})$ monomeric iron(II) chloride, cobalt(II) chloride and nickel(II) chloride molecules, there are also dimers (Fe_2Cl_4, Co_2Cl_4 and Ni_2Cl_4) present in the vapours with a halogen bridge structure:

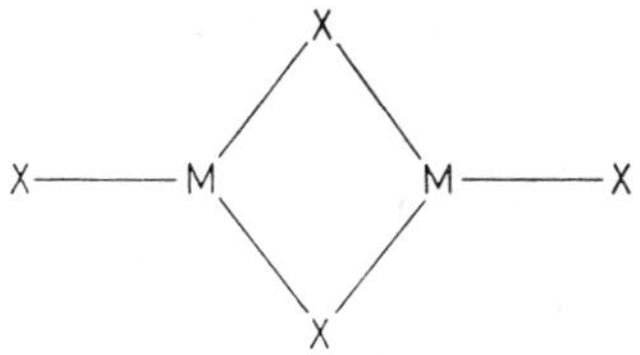

Bridge structures were shown to be consistent with matrix isolation Raman spectroscopic data for dimeric selenium dioxide (Boal, Briggs, Huber, Ozin, Robinson and Vander Voet, 1971; Ozin and Vander Voet, 1971):

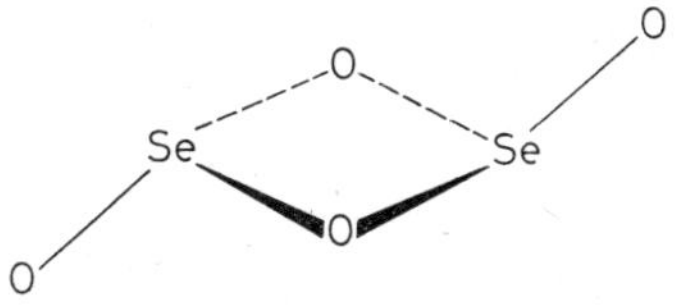

and dimeric germanium difluoride (Hastie, Hauge and Margrave, 1968; Zmbov, Hastie, Hauge and Margrave, 1968; Ozin, 1971):

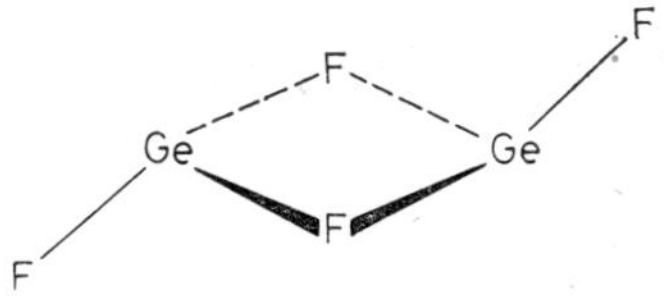

On the basis of matrix isolation spectroscopic and stereochemical data, the following configurations were established for the addition compounds $Cl_3P.N(CH_3)_3$* (Boal and Ozin 1972), and $Cl_4Si.N(CH_3)_3$ (Beattie and Ozin, 1970):

* The loop symbolizes the lone pairs of electrons.

CH_3, CH_3, CH_3 — N — Cl, Cl — P — Cl, Cl

CH_3, CH_3, CH_3 — N — Cl, Cl — Si — Cl, Cl, Cl

NON-RIGIDITY OF MOLECULAR CONFIGURATIONS

We conclude this survey of physical techniques for the determination of molecular geometry by mentioning an important stereochemical phenomenon. In both the elucidation and discussion of molecular geometries, it must be kept in mind that the molecules are not rigid bodies. The accurate determination of internuclear distances necessitates careful consideration of the molecular vibrations. Moreover, the establishment of the molecular shape and symmetry is also influenced by the intramolecular motion. The vibrational motions may permute the positions of the nuclei in the molecule, leading to stereoisomerisation (Muetterties, 1972). Molecular species that undergo such rearrangements are called *stereochemically non-rigid* molecules (Muetterties, 1965, 1970). The term "permutational isomerism" is also used (see, for example, Ugi and Ramirez, 1972). In general, stereochemical non-rigidity may refer to a certain physical technique if the permutational process is not slow compared with the time scale of the phenomenon on which the experiment is based (Muetterties, 1965). Ideally, the structures of non-rigid systems should be treated by means of the potential energy surface comprising all the possible configurations (Liehr, 1963). Not only the molecular configuration but even the coordination number observed may depend on the physical technique utilized in studying stereochemically non-rigid structures (Downs, 1968). The lifetimes of the ground vibrational states of some non-rigid systems are given in Table 2, and the time scales of the physical techniques are listed in Table 3. In connection with the unusual molecular structure of bis(cyclopentadienyl)beryllium dis-

TABLE 2

*Ground-state lifetimes for stereochemically
non-rigid molecules, from
Muetterties (1965)**

Molecule or ion	Lifetime, s
NH_3	2.5×10^{-11}
ND_3	2.7×10^{-10}
PH_3	10^{-3}
AsH_3	10
CH_4	10^{15}
PF_5	10^{-5}
PCl_5	10^4
PF_3Cl_2 $(-50\,^\circ C)$	10^{-3}
IF_7	10^{-3}–10^{-12}
$ReH_9{}^{2-}$	10^{-3}–10^{-12}
$Mo(CN)_8{}^{4-}$	10^{-3}–10^{-12}
$PF_4N(CH_3)_2$	10^{-4}
B_2Cl_4	10^{-3}–10^{-12}

* For references see Muetterties (1965).

TABLE 3

*Time scales of physical techniques for structure determination,
after Muetterties (1965)*

Technique	Approximate time scale, s
Electron diffraction	10^{-20}
Neutron diffraction	10^{-18}
X-ray diffraction	10^{-18}
Ultraviolet and visible spectroscopy	10^{-15}–10^{-14}
Infrared and Raman spectroscopy	10^{-13}
Electron spin resonance	10^{-4}–10^{-8}*
Nuclear magnetic resonance	10^{-1}–10^{-9}*
Mössbauer spectroscopy (iron)	10^{-7}
Molecular beam study	10^{-6}

* Sensitive to the system.

31

cussed earlier, it is interesting to note that the average lifetime of the beryllium atom in one of the two alternative positions has been estimated to be about 10^{-13}–10^{-12} s (Ionov and Ionova, 1970).

THE REPRESENTATIONS OF
MOLECULAR GEOMETRY*

The most complete and unambiguous description of the molecular geometry would be the r_e *equilibrium structure* characterized by the r_e equilibrium internuclear distances. The parameter r_e gives the distance between two hypothetically motionless nuclei in the free molecule and corresponds to the minimum of the potential energy function. It is this structure that is calculated by various semi-empirical methods (see, *e.g.*, Pople and Beveridge, 1970; Dewar, 1969) and the *ab initio* methods (see, *e.g.*, Pilar, 1968). The r_e structure could be determined experimentally only for some of the simplest molecules, however. The various internuclear distance parameters (r_0, r_s, r_a, etc.) obtained by the different physical techniques differ from the equilibrium values as a consequence of the vibrational and rotational motion of the molecule. If the equilibrium structure is not attainable, the *average structures* are the best representations of the molecular geometry. They have clear physical significance and can be obtained from the experimental data. The following internuclear distance parameters characterize the most important average structures:

r_α^0 (or r_z)   — the distance between average nuclear positions for the ground vibrational state;

r_v — the distance between average nuclear positions for the excited vibrational state, v;

r_α — the distance between average nuclear positions for thermal equilibrium;

r_g (or $r_g(0)$) — the thermal average value of the internuclear distance corresponding to the centre of gravity position of the probability distribution function.

* For references see, for example, Kuchitsu and Cyvin (1972).

32

Some of the important relationships between the distance parameters are given below:

$$r_g = r_e + \delta r + \langle \Delta z \rangle + \frac{\langle (\Delta x)^2 \rangle + \langle (\Delta y)^2 \rangle}{2r_e} + \ldots$$

$$r_\alpha = r_g - \delta r - \frac{\langle (\Delta x)^2 \rangle + \langle (\Delta y)^2 \rangle}{2r_e} + \ldots$$

or

$$r_\alpha = r_e + \langle \Delta z \rangle + \ldots$$

and

$$r_\alpha^0 = \lim_{T \to 0} r_\alpha.$$

These relationships refer to a Cartesian coordinate system whose z axis coincides with the equilibrium internuclear axis, and whose origin is the equilibrium position of the ith nucleus. Accordingly, the positions of the ith and jth nuclei are characterized as follows:

	Equilibrium	Instantaneous	Average
i	0 0 0	$\Delta x_i\, \Delta y_i\, \Delta z_i$	$\langle \Delta x_i \rangle \langle \Delta y_i \rangle \langle \Delta z_i \rangle$
j	0 0 r_e	$\Delta x_j\, \Delta y_j\, r_e + \Delta z_j$	$\langle \Delta x_j \rangle \langle \Delta y_j \rangle\, r_e + \langle \Delta z_j \rangle$
Difference	0 0 r_e	$\Delta x\, \Delta y\, r_e + \Delta z$	$\langle \Delta x \rangle \langle \Delta y \rangle\, r_e + \langle \Delta z \rangle$

The term $\langle \Delta z \rangle$ is the mean parallel amplitude and represents the deviation of the distance between the average positions of the nuclei due to anharmonicity. The quantities $\langle (\Delta x)^2 \rangle$ and $\langle (\Delta y)^2 \rangle$ are the mean square perpendicular amplitudes and δr represents the centrifugal distortion due to the rotation of the molecule.

The thermal average value of the internuclear distance, r_g, is related to the distance parameter, r_a, by the following expression:

$$r_g = r_a + \frac{l^2}{r_e}$$

where l^2 or $\langle (\Delta z)^2 \rangle$ is the mean square amplitude of vibration. The correction term contains the distance parameter, r_e. However, which of the distance parameters is used in the correction term is of negligible importance, and it is also not of significance which of the l values is

3

used in this correction. For example, $l_g^2 = \langle (r - r_g)^2 \rangle$ corresponds to the r_g structure. Utilizing the relationships presented above, the electron diffraction results can be converted into the r_α^0 structure comprising the distances between the average nuclear positions in the ground vibrational state.

As for the other principal technique, internuclear distance parameters with well defined physical meaning cannot be obtained from either the r_0 or the r_s structures. However, as was mentioned, the r_s parameters are very similar to the r_e equilibrium values, more so than the r_0 parameters. This point can be seen from the approximate relationship $r_s = (r_0 + r_e)/2$.

Average structures with a well defined physical meaning can be obtained by considering the relationship between the effective and equilibrium values of the rotational constants:

$$A_0 = A_e + \text{vibrational–rotational corrections.}$$

The vibrational–rotational corrections consist of two parts: a harmonic and an anharmonic part. The harmonic correction can be evaluated from a knowledge of the harmonic force field of the molecule. Thus the A_z rotational constants are obtained:

$$A_z = A_0 + \text{harmonic correction}$$

corresponding to the molecular geometry characterized by the r_z average internuclear distances. The r_z structure derived from the microwave spectroscopic data is essentially identical with the r_α^0 structure attainable from the electron diffraction analysis. Comparison of the results in terms of r_α^0 and r_z structures is then the best means of rationalizing the structural information obtained by the two techniques. The r_z (r_α^0) structure together with the r_g distance parameters are the best representations of the molecular geometry.

Clear physical significance and experimental attainability are the two most important merits of the representations in terms of r_z (r_α^0) and r_g structures (Kuchitsu, 1968). The distance parameter, r_g, is the most convenient means of characterizing the average length of a chemical bond. It has no such descriptive geometrical meaning, however, for distances between non-bonded atoms because of the effects of perpendicular vibrations. Accordingly, the most suitable represen-

34

tation of the complete geometry of a molecule is the r_z (r_α^0) structure. Instead of a real average bond distance, however, this representation yields the average projection of the bond on to the line joining the equilibrium nuclear positions ($r_e + \langle \Delta z \rangle$).

The bond angles* calculated from the r_e or r_z (r_α^0) representations of the internuclear distances are well defined, as they are the equilibrium and zero-point average bond angles, respectively. The bond angle corresponding to the r_α structure refers to the thermal average nuclear positions. The bond angles calculated from the substitution coordinates obtained by microwave spectroscopy have no such well defined significance but are very similar to the equilibrium bond angles. When no specifications and corrections are indicated, the electron diffraction studies usually provide bond angles calculated from the r_a distances. In some instances the angles calculated from the r_g distances are given. These angles are not, however, the thermal average values. It is considered to be more convenient to represent the geometry as r_g values for the bond distances and the bond angles (and torsional angles) as calculated from an r_α structure. This representation is called the r_y structure and has yet to become generally applied (Bartell, Kuchitsu and Seip, 1974).

The differences between various representations are seldom greater than the experimental uncertainties. This aspect is, however, gradually gaining importance with the improvement of experimental and computational techniques.

The question of the representation of the molecular geometry is certainly important, however, for molecules that undergo large amplitude motion. In such instances careful consideration is required as important features of the molecular geometry may be concealed by the effects of molecular vibrations. In this respect, the term $[\langle (\Delta x)^2 \rangle + \langle (\Delta y)^2 \rangle]/(2r_e)$ is the most significant in the correction $r_g - r_\alpha^0$. Often the notation

$$K = \frac{\langle (\Delta x)^2 \rangle + \langle (\Delta y)^2 \rangle}{2r_e}$$

is used. An extreme example is given here for illustration. A nozzle temperature of about 800°C was used in electron diffraction experi-

---

* For detailed discussion and references, see Kuchitsu (1971).

ments with manganese(II) chloride (Hargittai, Tremmel and Schultz, 1975) and the following internuclear distances were obtained:

$$r_g \, (Mn-Cl) \; = \; 2.202 \pm 0.005 \; \text{Å}$$
$$r_g (Cl \ldots Cl) \; = \; 4.319 \pm 0.015 \; \text{Å}$$

corresponding to a bond angle of 155°. On the other hand, it is known from molecular beam deflection studies (Büchler, Stauffer and Klemperer, 1964b) that the $MnCl_2$ molecule has a linear average configuration. The difference

$$2r_g(Mn-Cl) - r_g(Cl \ldots Cl) = 0.085 \; \text{Å}$$

called the *shrinkage**, is a consequence of perpendicular vibrations. The r_α^0 structure (and, in fact, the r_α structure) is free from this effect.

The simplest case of the shrinkage effect was demonstrated by the example of the triatomic linear molecule. This effect is usually much decreased when the experiments are performed at ordinary temperatures but it should never be excluded from consideration, even for complicated molecules, if large-amplitude motion is present. The shrinkage is usually denoted by δ (δ_g and δ_a for the r_g and r_a structures, respectively) and is obtained from the spectroscopic calculations by the K value.

As the linkages between different parts of coordination molecules may often be loose, motions of large amplitude probably occur and, accordingly, careful examination of the influence of the intramolecular motion in establishing the molecular geometry is desirable. This again stresses the importance of correlation between the mean amplitudes of vibration (l values) and the geometrical parameters and, consequently, the bonding properties of the coordination molecules. Comparison of the l values determined for the complexes with those observed for the free (uncomplexed) molecules is particularly relevant. Studies referring to such comparisons have been initiated (Cyvin, 1973a).

*

* It is also called the Bastiansen–Morino shrinkage effect; for references, see Cyvin (1968).

A final note concerns the uncertainties in the structural parameters indicated in our compilations. With one or two exceptions, the values as given by the original authors are reproduced. The origins and meaning of these data, however, may be different. Accordingly, for further use of the structural parameters, particularly when studying effects whose magnitudes are comparable with the experimental error, it is necessary to consult the original papers for more detailed information. For the electron diffraction studies, special mention is made if the work was performed by the visual technique. Very careful handling of such data is called for because, apart from the large uncertainties in the molecular parameters, the molecular configurations established have proved to be incorrect in several instances. Many of the studies with the visual technique, however, withstand comparison with up-to-date methods as far as the molecular geometry is concerned. In many instances where we feel it to be of importance, the assumptions used in the structural analyses are stated. The microwave spectroscopic and electron diffraction findings are most often reported in terms of r_0 and $r_a = r_g(1)$ parameters, respectively; this is not always mentioned. If any other representations of the molecular geometry are given, special mention is always made.

ADDITION COMPOUNDS

In addition compounds there are two partner molecules, one being an electron donor (Lewis base) and the other an electron acceptor (Lewis acid). Thus the term "addition compound" is appropriate. These compounds are sometimes also referred to as donor–acceptor complexes or charge–transfer complexes. The linkage between the two partners is stronger than the van der Waals interaction.

The molecules $R_3Q.ZX_3$ (Q = N or P, Z = B or Al, R the ligand in the donor and X the ligand in the acceptor) will be considered in this chapter, although the properties described above may character- ize, to various extents, all coordination compounds.

In the molecules $R_3N.BX_3$, for instance, the nitrogen with its lone pair of electrons is the electron donor and the boron atom with its vacant orbital is the electron acceptor. The linkage is formed as the lone pair of electrons on the nitrogen atom is shared between the nitrogen and boron atoms. The stability of the addition compounds is determined by both the donor strength and the acceptor strength of the linked molecules. A detailed theoretical discussion illustrated by experimental data (which are mainly spectroscopic) is given in the book by Mulliken and Person (1969).

This chapter focuses on the results of vapour-phase structural deter- minations of boron–nitrogen, boron–phosphorus and aluminium– nitrogen compounds. Relevant crystal-phase data are also included.

In addition to reviewing the determinations of the molecular geom- etries of individual compounds, the variations of the geometrical parameters in series of compounds and possible correlations between geometry and bonding are also considered.

The following two main points are examined in detail: (1) the molec- ular geometries of the donor and acceptor parts compared with the structure of the monomers; (2) the variations in the complex geometry and especially the donor–acceptor bond distance in the presence of various substituents.

The results of structural determinations on monomers are considered first.

MONOMERS

The internuclear distances in BF_3 and BCl_3 determined by up-to-date methods are presented in Table 4. In addition to the importance of the numerical results, it can be seen from the ratios of the $r(X \ldots X)$ and $r(B-X)$ values, which are $\sqrt{3}$ within the error limits, that the average structures are trigonal planar. This also applies to the equilibrium configuration. Although this is to be expected, there are not

TABLE 4

Internuclear distance parameters for BX_3 molecules

		$r(B-X)$, Å		$r(X \ldots X)$, Å	
BF_3	r_g	1.3156 ± 0.0044	(a)	2.2733 ± 0.0041	(a)
	r_α	1.3130 ± 0.0044	(b)	2.2723 ± 0.0041	(b)
	r_a	$1.311_9 \pm 0.0008$	(c)	$2.270_5 \pm 0.001_3$	(c)
	r_g	$1.313_3 \pm 0.001_0$	(c)		
	r_α	1.310_9	(c)		
	r_α^0	$1.311_1 \pm 0.001_2$	(c)		
	r_0	1.309_5	(d)		
	r_z	$1.311_2 \pm 0.001$	(e)		
BCl_3	r_g	1.7421 ± 0.0044	(a)	3.0134 ± 0.0060	(a)
	r_α	1.7387 ± 0.0044	(b)	3.0119 ± 0.0060	(b)
BBr_3	r_g	1.8932 ± 0.0054	(f)	3.2830 ± 0.0053	(f)
$B(CH_3)_3$	r_g	$1.578_3 \pm 0.001_1$	(g)	$2.725_1 \pm 0.002_7$	(g)

(a) Electron diffraction, Konaka, Murata, Kuchitsu and Morino (1966). The non-linear shrinkage effects corresponding to the planar configuration are 0.005 Å and 0.004 Å for BF_3 and BCl_3, respectively.

(b) Electron diffraction, Konaka, Murata, Kuchitsu and Morino (1966), using the correction terms (cf., p. 33).

$$r_g - r_\alpha = \frac{\langle (\Delta x)^2 \rangle + \langle (\Delta y)^2 \rangle}{2r} + \delta r \, .$$

The non-linear shrinkage effect vanishes in the r_α structure. There is some discrepancy between the electron diffraction data and earlier spectroscopic results of Nielsen (1954).

(c) Results of an electron diffraction reinvestigation by Kuchitsu and Konaka (1966), $\delta_a = \sqrt{3} \; r_a(B-F) - r_a(F \ldots F) = 0.001_7 \pm 0.001_1$ Å.

(d) Determined from the rotational structure of the infrared spectrum measured by Nielsen (1954) and corrected by Kuchitsu and Konaka (1966).

(e) Ground-state average structure value corresponding to the electron diffraction and spectroscopic data.

(f) Electron diffraction, Konaka, Ito and Morino (1966).

(g) Electron diffraction, Bartell and Carroll (1965b), $<C-B-C = 119.4 \pm 0.3°$ with the shrinkage neglected, and assuming an equilibrium bond angle of 120.0°.

TABLE 5

Molecular geometries of ammonia and its methyl derivatives

	NH_3	NH_2CH_3	$NH(CH_3)_2$	$N(CH_3)_3$
$r(N-H)$, Å	1.019±0.002 (a) 1.024±0.001 (b) 1.030±0.002 (c) 1.022$_7$±0.003 (d) 1.0173 (e)	1.011 (f)	1.00 ±0.02 (h)	
$r(N-C)$, Å		1.465±0.002 (g) 1.474 (f)	1.455±0.002 (h)	1.454±0.002 (h) 1.451±0.003 (i)
<H−N−H	109.1±1.0° (a) 107.3$_2$° (b) 108.$_2$±1.1° (d) 107.78° (e)	105°52′ (f)		
<C−N−H		112°3′ (f)	107±2° (h)	
<C−N−C			111.8±0.6° (h)	110.6±0.6° (h)

(a) Electron diffraction, r_a parameters, Bastiansen and Beagley (1964). The same paper also contains the results of an investigation on ND_3.

(b) r_z structure obtained from the rotational constants, Morino, Kuchitsu and Yamamoto (1968).

(c) Electron diffraction, r_g parameters, Kuchitsu, Guillory and Bartell (1968).

(d) r_a^0 structure, Kuchitsu, Guillory and Bartell (1968). The same paper also contains the results of an investigation on ND_3.

(e) r_0 structure from the rotational spectrum, Benedict and Plyler (1957).

(f) Microwave spectroscopy, Lide (1957).

(g) Electron diffraction, r_g parameters, Higginbotham and Bartell (1965). The structure of ND_2CD_3 was also determined.

(h) Electron diffraction, $r_g(1)$ parameters, Beagley and Hewitt (1968).

(i) Microwave spectroscopy, Wollrab and Laurie (1969).

TABLE 6

Vapour-phase molecular geometries of phosphine and its simple derivatives

	r(P–H), Å	r(P–C), Å	<H–P–C <H–P–F	<C–P–C	r(P–F), Å	<F–P–F
PH$_3$	1.437 ± 0.004 (a)					
PH$_2$CH$_3$	1.423 ± 0.007 (b)	1.858 ± 0.003 (b)	96.5°* (b)			
P(CH$_3$)$_3$		1.846$_5$ ± 0.003 (c) 1.841 ± 0.003 (d)		98.6 ± 0.3° (c) 99.1 ± 0.2° (d)		
PF$_3$					1.568$_9$ ± 0.001$_2$ (e) 1.564$_7$ ± 0.001$_2$ (f) 1.537 ± 0.004 (g)	97.8 ± 0.2° (f) 98.2° (g)
PHF$_2$	1.412 ± 0.006 (h)		96.3 ± 0.5° (h)		1.582 ± 0.002 (h)	99.0 ± 0.2° (h)

* Assumed

(a) Electron diffraction, r_g parameters, Bartell and Hirst (1959). A reliable result could not be obtained for the bond angle. The value 1.419 Å was estimated for r_e.
(b) Electron diffraction, r_g parameters, Bartell (1960).
(c) Electron diffraction, r_g parameters, Bartell and Brockway (1960).

(d) Microwave spectroscopy, r_0 structure, Lide and Mann (1958).
(e) and (f) Electron diffraction, Morino, Kuchitsu and Moritani (1969). (e) r_a parameter, (f) r_α^0 structure.
(g) Hersch (1963).
(h) Microwave spectroscopy, Kuczkowski (1968).

many molecules, even among the simplest types, for which the equilibrium configuration has been actually determined and not just assumed.

The molecular geometries of ammonia and its methyl derivatives are compiled in Table 5 and the data for some simple phosphorus compounds are presented in Table 6. No further comment is felt to be necessary here.

Studies and results on the molecular geometries of monomeric aluminium derivatives are now discussed in more detail. As aluminium trihalides are evaporated, dimeric molecules are often formed. There is a pressure- and temperature-dependent equilibrium between the monomeric and dimeric species in the vapour phase.

Two maxima appeared at 1.63 and 2.82 Å on the radial distribution obtained from the electron scattering data for aluminium trifluoride. These values correspond to the $Al-F$ bond and $F \ldots F$ non-bond interaction of the planar monomeric AlF_3 molecule (Akishin, Rambidi and Zasorin, 1959). Similarly, it was easy to interpret the experimental data for aluminium triiodide (Table 7).

TABLE 7

Geometrical parameters for AlX_3 molecules

AlX_3		$r(Al-X)$, Å	$r(X \ldots X)$, Å
AlF_3	(a)	1.63 $\pm$ 0.01	2.82 $\pm$ 0.02
$AlCl_3$	(b)	2.06 $\pm$ 0.01	3.53 $\pm$ 0.01
AlI_3	(a)	2.44 $\pm$ 0.02	
	(c)	2.449 $\pm$ 0.013	
AlH_3	(d)	1.715 $\pm$ 0.010	
$Al(BH_4)_3$	(e)	1.801 $\pm$ 0.004	
$Al(CH_3)_3$	(f)	1.957 $\pm$ 0.003	3.390 $\pm$ 0.005

(a) Electron diffraction results, presumably r_m parameters. It was determined that $<F-Al-F = 120°$, while it was only assumed that $<I-Al-I = 120°$, Akishin, Rambidi and Zasorin (1959a).

(b) Electron diffraction, $r_g(1)$ parameters, the corresponding $<Cl-Al-Cl = 118 \pm 1.5°$, Zasorin and Rambidi (1967a).

(c) Shen (1973).

(d) In the crystalline phase, X-ray diffraction study, Turley and Rinn (1969).

(e) Electron diffraction, r_a parameter, bridge structure, Almenningen, Gundersen and Haaland (1968a). For a more detailed discussion, see chapter on borohydrides.

(f) Electron diffraction, r_a parameters. The assumption of 120° for the $C-Al-C$ corresponds to a shrinkage of 0.012 Å for the $C \ldots C$ distances, Almenningen, Halvorsen and Haaland (1971).

For aluminium chloride, there are mostly dimeric molecules in the vapour phase at the pressure necessary for the electron diffraction experiment. The reaction

$$Al_2Cl_6 \rightleftharpoons 2AlCl_3$$

can be shifted towards the dissociation of the dimeric species by a considerable increase in temperature. The temperature dependence of the equilibrium constant of this reaction has been studied in detail (Smits and Meijering, 1938; Vrieland and Stull, 1967). The vapour ot aluminium chloride consists mainly of monomeric $AlCl_3$ molecules a, a temperature of 800 K. The corresponding vapour pressure isf however, much too high for electron diffraction patterns to be recorded for evaluation. The main problem is that in the open effusion cell the sample would leave before reaching the desired temperature. The solution to this problem was found by Rambidi and Zasorin (1964), who used a so-called double effusion chamber, as shown in Fig. 3. The sample is placed in the lower chamber, which is heated to a temperature T_1, corresponding to the optimum conditions for the electron scattering experiment. The temperature of the upper chamber, $T_2 > T_1$, then determines the composition of the vapour leaving the nozzle tip. The temperature T_2 is chosen so as to ensure

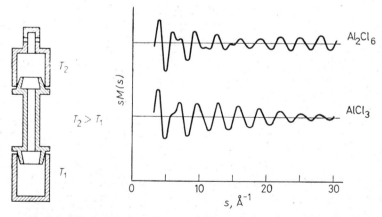

Fig. 3. Schematic diagram of the double effusion chamber (Rambidi and Zasorin, 1964)

Fig. 4. The electron scattering molecular intensities of monomeric and dimeric aluminium trichloride (Zasorin and Rambidi, 1967a)

43

that the vapour has a maximum content of the monomeric molecules. For this purpose, it is of great significance to design the geometry of the double chamber properly, taking the diffusion and dissociation processes into consideration.

Figure 4 demonstrates by means of the molecular electron scattering intensities that the electron diffraction pattern becomes simpler with increasing monomeric content in the vapour of aluminium chloride.

The molecular geometry of monomeric aluminium trichloride has been the subject of discussion in the recent literature. As the knowledge of the structure of the monomer is a key issue in interpreting variations in the geometries of the aluminium complexes, the evidence from various experimental studies is briefly surveyed below.

Zasorin and Rambidi (1967a) obtained internuclear distances from the electron diffraction data on $AlCl_3$ vapour at about 800 K of $r(Al-Cl) = 2.06 \pm 0.01$ Å and $r(Cl \ldots Cl) = 3.53 \pm 0.01$ Å, corresponding to a $Cl-Al-Cl$ bond angle of $118 \pm 1.5°$. These values were determined from the radial distribution curve and are $r_g(1)$ parameters. Although the apparent configuration was flat pyramidal, Zasorin and Rambidi (1967a) suggested that the deviation from planarity is a consequence of the shrinkage effect. The corresponding value of $\delta_g = 0.04$ Å for the shrinkage of the Cl . . . Cl distances was used in subsequent calculations (Zasorin and Rambidi, 1967b) to estimate a frequency value of 95 ± 15 cm^{-1} for the out-of-plane vibration (v_2) of $AlCl_3$. This study was, in fact, pioneering in demonstrating the possibility of the determination of vibrational frequencies from shrinkage values obtained by electron diffraction.

Although the shrinkage values usually cannot be determined from the electron diffraction data with sufficient accuracy for determining vibrational frequencies (Cyvin, Cyvin, Brunvoll, Andersen and Stølevik, 1967), the requirements for accuracy are considerably lower for high-temperature data (Cyvin and Brunvoll, 1969). This was amply demonstrated by the calculations on $AlCl_3$ by Cyvin and Brunvoll (1969) who obtained the value of 110 cm^{-1} for v_2, in satisfactory agreement with the findings of Zasorin and Rambidi (1967b). The curves in Fig. 5, expressing the relationship between the shrinkage effect and the v_2 frequency for $AlCl_3$, demonstrate convincingly the advantage of using high-temperature experiments for such studies. It can also be seen that the method is expected to be used successfully, particularly for low frequencies.

44

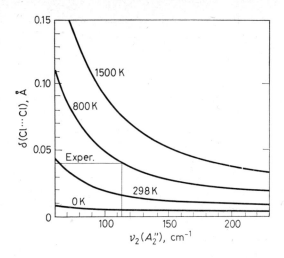

Fig. 5. The shrinkage effect, $\delta(Cl \ldots Cl)$ in $AlCl_3$ at different temperatures calculated as a function of the frequency (ν_2) of the out-of-plane motion (Cyvin and Brunvoll, 1969)

More recently, the bending vibrational frequencies of $CoCl_2$ (Tremmel, Ivanov, Schultz, Hargittai, Cyvin and Eriksson, 1973) and $MnCl_2$ (Hargittai, Tremmel and Schultz, 1975) have been obtained from the shrinkages determined by high-temperature electron diffraction experiments. Indeed, the study on manganese chloride was the first case when experimental evidence (molecular beam deflection investigation of polarity by Büchler, Stauffer and Klemperer, 1964) was available on the basis of which an assumption could be made concerning the equilibrium molecular configuration. For both $AlCl_3$ and $CoCl_2$, such experimental data were not available.

The assumption of the planar equilibrium configuration (D_{3h} symmetry) of the $AlCl_3$ molecule seems to be reasonable, and is also supported by the experimental evidence for the planar configuration of the BCl_3 molecule. However, caution is necessary because there are various discrepancies between the bonding of boron and aluminium compounds (cf., Lesiecki and Shirk, 1972). There are precedents for changes in the molecular configurations as the atomic number of the central atom changes in other groups of the periodic system.

The vibrational spectra of monomeric aluminium chloride also provide important data for establishing its molecular configuration. The compilation of gas-phase and matrix isolation vibrational frequencies by Lesiecki and Shirk (1972) is reproduced in Table 8 and is augmented by the more recent data from a xenon matrix study by Perov (1973). The far-infrared spectrum of $AlCl_3$ obtained in the argon matrix is shown in Fig. 6. On the basis of the vibrational spectra, the most important conclusions could be made from the occurrence of the symmetric stretching frequency (v_1) and from the value of the v_2 frequency. If the symmetric stretching is infrared-active, a pyramidal configuration with C_{3v} symmetry would be indicated. The values assigned to v_2 in the two independent experimental studies are in excellent agreement. This frequency was predicted by Selivanov and Mal'tsev (1973) to be 160–170 cm^{-1} on the basis of the data for other trihalides. The calculated values, accordingly, seem to be rather low. Because of the importance of this question, further studies are expected to be carried out. It is only fair to state, however, that on the basis of the available electron diffraction data it is not possible to decide whether the $AlCl_3$ molecule has a flat pyramidal or a planar average structure. Moreover, electron diffraction alone will not be capable of distinguishing between such models.

Concerning the assignment of $v_2 = 182.8$ cm^{-1} by Lesiecki and Shirk (1972), the authors supplemented their study with a normal coordinate analysis which suggested a $Cl-Al-Cl$ angle near 112°. This indicates a rather peaked pyramidal configuration, and an inver-

TABLE 8

Vibrational frequencies (cm^{-1}) of $AlCl_3$

	In argon matrix (a)	In xenon matrix (b)	In gaseous phase	
			Raman (c)	Infrared (d)
v_1	382.2		375	
v_2	182.8	174		
v_3	594.7			610
v_4	149.2	142	150	

(a) Lesiecki and Shirk (1972).
(b) Perov (1973).
(c) At 650°C, Beattie and Horder (1969).
(d) At 900°C, Klemperer (1956).

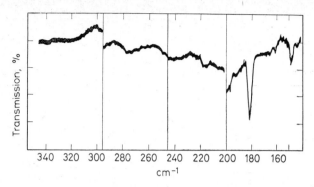

Fig. 6. The far-infrared spectrum of AlCl₃ in argon matrix
(Lesiecki and Shirk, 1972)

sion barrier that is 2–3 vibrational quanta high. Hargittai and Hargittai (1974) showed, however, that the electron diffraction data (Zasorin and Rambidi, 1967) are not compatible with such an interpretation. According to the molecular orbital calculations by Companion (1972), a donor–acceptor complex is probably formed in the argon matrix of $AlCl_3$, which may be connected with the occurrence of a frequency of 182.8 cm^{-1}.

The geometrical parameters of the AlX₃ monomers (X = F, Cl, I, H, BH_4 or CH_3) are collected in Table 7.

BORON–NITROGEN COMPLEXES

The molecular geometries of only a few boron–nitrogen donor–acceptor complexes have been determined so far. Vapour-phase investigations are often hindered by the low volatility of these compounds and their tendency to dissociate. The molecules studied so far are listed in Table 9, together with the lengths of the boron–nitrogen bonds and the corresponding distances calculated for the covalent and van der Waals linkages. In obtaining the calculated values, the covalent and van der Waals radii used were taken mainly from Pauling (1960).

Before making more general comments on the molecular structures of the boron–nitrogen complexes, the data referring to the individual studies will be considered.

47

TABLE 9

B—N bond lengths (Å) *in boron–nitrogen addition molecules*

Compound	$r(B-N)$, gas	$r(B-N)$, crystal	$\Sigma r_{cov.}$*	$\Sigma r_{v.d.w.}$*
$NH_3.BF_3$		1.60 ± 0.015 (c1)	1.51	3.0
$CH_3NH_2.BF_3$		1.57 (c2)		
$(CH_3)_3N.BF_3$	1.636 ± 0.004 (a1)	1.585 (c3)		
$(CH_3)_3N.BCl_3$		1.575 ± 0.011 (c4)		
		1.610 ± 0.006 (c5)		
$(CH_3)_3N.BBr_3$		1.603 ± 0.02 (c4)		
$(CH_3)_3N.BI_3$		1.584 ± 0.025 (c4)		
$(CH_3)_3N.BH_3$	1.62 ± 0.15 (b)			
	$1.65_5 \pm 0.02$ (a2)			
	1.637 (a3)			
	1.609			
$CH_3CN.BF_3$		1.635 (c6)		
		1.630 ± 0.004 (c7)		
$CH_3CN.BCl_3$		1.562 ± 0.008 (c7)		
$(CH_3)_3N.B(CH_3)_3$	1.70–1.95 (a4)			
$C_5H_5N.BF_3$		1.53 (c8)		

(a) Microwave spectroscopy, (1) Bryan and Kuczkowski (1971), (2) Schirdewahn (1965), (3) Durig, Li and Odom (1973), (4) Lide, Taft and Love (1959).

(b) Visual technique of electron diffraction, Bauer (1937).

(c) Crystalline phase X-ray diffraction, (1) Hoard, Geller and Cashin (1951), (2) Geller an d Hoard (1950), (3) Geller and Hoard (1951), (4) Clippard, Hanson and Taylor (1971), (5) Hes s (1969), (6) Hoard, Owen, Buzzell and Salmon (1950), (7) Swanson, Shriver and Ibers (1969) (8) Zvonkova (1956).

* $r_{cov.}$ = covalent radii; $r_{v.d.w.}$ = van der Waals radii, Pauling (1960).

The microwave spectra of three isotopic species of *trimethylamine-boron trifluoride*, $(CH_3)_3N.BF_3$, have been recorded by Bryan and Kuczkowski (1971). The value of $r(B-N)$ could be determined reliably as the boron and nitrogen atoms lie on the symmetry axis. However, the other geometrical parameters could not be evaluated.

The results of two independent microwave spectroscopic investigations on *trimethylamine–borane*, $(CH_3)_3N.BH_3$, by Schirdewahn (1965) and Durig, Li and Odom (1973) are collected in Table 10. The data for the BH_3 group are the most accurate as they correspond to the r_s structure. The other parameters, except $r(C-N)$, were

TABLE 10

Geometrical parameters for trimethylamine–borane determined by microwave spectroscopy

	(a)	(b)	(c)
$r(C-H)$, Å	1.09 (assumed)	1.111	1.095 (assumed)
$r(B-H)$, Å	1.17 (assumed)	1.211 ± 0.003	1.211 ± 0.003
$r(B-N)$, Å	$1.65_5 \pm 0.02$	1.637	1.609
$r(C-N)$, Å	1.490 ± 0.005	1.495	1.495
$<H-B-H$	$112.2 \pm 2.0°$		
$<N-B-H$		$105.32 \pm 0.16°$	$105.32 \pm 0.16°$
$<B-N-C$		109.6°	110.9°
$<C-N-C$	$110.1 \pm 1.0°$		
$<H-C-H$	$110.8 \pm 0.5°$		
$<N-C-H$		106.6°	110.1° (assumed)

(a) Schirdewahn (1965).
(b) and (c) Durig, Li and Odom (1973). The values for $r(B-H)$ and $<N-B-H$ are r_s parameters. For obtaining the other parameters different assumptions have been used in (b) and (c), respectively.

sensitive to the assumptions used in fitting the parameters to the rotational constants. The determination of $r(B-N)$, the most interesting parameter, is rendered more difficult by the nitrogen atom being near the centre of mass of the molecule. It is relevant here to mention the crystal-phase X-ray determination of the molecular structure of *hexamethylenetetramine–borane*, $C_6H_{12}N_4.BH_3$, because part of this molecule is analogous to $(CH_3)_3N.BH_3$ (Hanic and Subrtová, 1969). The parameters that may be useful for comparison are given in Fig. 7.

The microwave spectrum of only a single isotopic species was analyzed for *trimethylamine–trimethylborane*, $(CH_3)_3N.B(CH_3)_3$, by Lide, Taft and Love (1959). Therefore, not much information could be expected from the experimental data, i.e., one moment of inertia, because at least nine independent structural parameters would be required to determine the molecular geometry of this compound. Certain conclusions could be made, however, by using various assumptions. When the trimethylamine and trimethylborane geometries in the complex were assumed to be the same as in the respective monomers, a value of 1.80 Å for the length of the $B-N$ bond was calculated by Lide, Taft and Love (1959). This parameter proved to be less

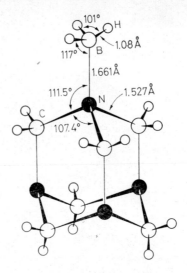

Fig. 7. The crystal-phase molecular structure of hexamethylenetetramine–borane with the geometrical parameters of the $H_3B.NC_3$ fragment (Haňic and Subrtová, 1969)

sensitive to the assumptions made for the bond distances than those for the bond angles. Allowing an uncertainty of $\pm 4°$ for both angles and assuming that they are not appreciably smaller than tetrahedral, it was concluded that the interval of 1.70 to 1.95 Å is the best estimate for the value of $r(B-N)$.

More studies have been performed on crystalline boron–nitrogen addition compounds than on the vapour-phase compounds. The data for the $B-N$ bond length are given in Table 9. Unfortunately, most of the uncertainties in the parameters determined are large, partly because some of the studies were performed many years ago with less modern techniques. Another source of uncertainty is the particular characteristic of many addition compounds that their crystals are often difficult to handle. Nevertheless, the data on the crystalline molecular structure of boron–nitrogen addition compounds are very valuable in discussing the structural properties of the boron–nitrogen complexes. However, one should keep in mind their uncertainty and crystalline state. As will be seen, the variations in the geometrical data within small series of boron–nitrogen complexes are consistent, with certain trends apparent. Contradictions between these trends, however, hinder the shaping of a more general pattern of relationships among these data.

The structural results on $R_3N.BF_3$ complexes collected in Table 11 seem to correlate well with the data referring to the relative stabilities of the compounds in question (Hoard, Geller and Owen, 1951). Definite conclusions cannot be drawn, however, because of the controversial character of many data on addition compounds (cf., p. 61.). In crystalline boron–nitrogen complexes, as expected, the longest $B-N$ bond appears in the least stable compound, $CH_3CN.BF_3$, indicating the weakness of the linkage. Further, the shortest $B-N$ bond was observed in the most stable compound $(CH_3)_3N.BF_3$. A gas-phase study on the same compound, however, indicated a much longer $B-N$ distance, about as long, in fact, as that in the crystalline $CH_3CN.BF_3$. The origin of the discrepancy may lie either in the structural difference between the gas-phase and crystal-phase molecules, or in the uncertainties in the X-ray diffraction determinations.

There is a marked, although not significant, difference in the values of $r(B-N)$ obtained for $(CH_3)_3N.BCl_3$ by two independent X-ray diffraction studies. The data in Table 9 show a gradual shortening of the $B-N$ bonds in the molecules $(CH_3)_3N.BX_3$ as X changes from chlorine towards iodine. This trend indicates the strengthening of the acceptor properties of the BX_3 part towards trimethylamine.

A similar trend was observed by Swanson, Shriver and Ibers (1969) for crystalline $CH_3CN.BF_3$ and $CH_3CN.BCl_3$, where the $B-N$ distance was found to be 0.068 ± 0.009 Å shorter for the BCl_3 complex. Normal coordinate vibrational analyses showed (Swanson and Shriver, 1970; Shriver and Swanson, 1971) that the $B-N$ stretching force constant for the BF_3 complex is smaller (2.5 mdyn/Å) than for the BCl_3 derivative (3.4 mdyn/Å), thus indicating a significantly stronger linkage for BCl_3 than for BF_3. It is of interest to mention that the BCl_3 complex is more stable than the BF_3 complex. Considerations on the reorganization of the boron halides may be helpful in interpreting the above variations (Swanson, Shriver and Ibers, 1969). In fact, the $Cl-B-Cl$ angle is smaller in $CH_3CN.BCl_3$ than the $F-B-F$ angle in $CH_3CN.BF_3$ (see Table 11). It is also observed that the boron–halogen stretching is more extensive in the complex formation for BCl_3 than for BF_3. It has been suggested that the partial

TABLE 11

Crystallographic results for boron–nitrogen and boron–phosphorus addition compounds

Early studies on four boron–nitrogen compounds

	$NH_3.BF_3$ (a)	$CH_3NH_2.BF_3$ (b)	$(CH_3)_3N.BF_3$ (c)	$CH_3CN.BF_3$ (d)
$r(B-N)$, Å	1.60	1.57	1.585	1.635
$r(B-F)$, Å	1.38	1.39	1.39	1.33
$r(C-N)$, Å		1.50	1.50	
$<F-B-F$	111°	110.5°	107°	114°
$<F-B-N$	107°	108.5°	112°	103°
$<B-N-C$		114°	105°	
$<C-N-C$			114°	

(a) Hoard, Geller and Cashin (1951).
(b) Geller and Hoard (1950).
(c) Geller and Hoard (1951).
(d) Hoard, Owen, Buzzell and Salmon (1950).

$CH_3CN.BX_3$ (X = F, Cl), Swanson, Shriver and Ibers (1969)

	$CH_3CN.BF_3$	$CH_3CN.BCl_3$
$r(B-N)$, Å	1.630 ± 0.004	1.562 ± 0.008
$r(B-X)$, Å	1.343 ± 0.004	1.825 ± 0.005
$r(C-N)$, Å	1.135 ± 0.005	1.122 ± 0.007
$r(C-C)$, Å	1.439 ± 0.005	1.437 ± 0.008
$<X-B-X$	$112.7\pm0.2°$	$111.9\pm0.2°$
$<X-B-N$	$105.5\pm0.3°$	$106.8\pm0.4°$
$<C-N-B$	$179.0\pm0.3°$	$178.4\pm0.6°$
$<C-C-N$	$179.3\pm0.4°$	$179.7\pm0.6°$

$(CH_3)_3N.BX_3$ (X = Cl, Br, I), Clippard, Hanson and Taylor (1971)

	$(CH_3)_3N.BCl_3$	$(CH_3)_3N.BBr_3$	$(CH_3)_3N.BI_3$
$r(B-N)$, Å	1.609 ± 0.006	1.60 ± 0.02	1.58 ± 0.03
$r(B-X)$, Å	1.864 ± 0.004	2.04 ± 0.02	2.28 ± 0.02
$r(C-N)$, Å	1.52 ± 0.02	1.52 ± 0.02	1.52 ± 0.02
$<X-B-X$	$109.4\pm0.2°$	$108.3\pm0.6°$	$108.2\pm0.6°$
$<X-B-N$	$109.3\pm0.3°$	$110.1\pm0.7°$	$110.7\pm0.8°$
$<C-N-C$	$108.6\pm0.3°$	$107.6\pm1.0°$	$105.0\pm1.5°$
$<C-N-B$	$110.8\pm0.2°$	$111.3\pm0.8°$	$112.8\pm1.2°$

TABLE 11 (continued)

$(CH_3)_3P.BX_3$ $(X = Cl, Br, I)$, Black (1971)

	$(CH_3)_3P.BCl_3$	$(CH_3)_3P.BBr_3$	$(CH_3)_3P.BI_3$
$r(B-P)$, Å	1.957	1.924	1.918
$r(B-X)$, Å	1.855	2.022	2.249
$r(C-P)$, Å	1.817	1.812	1.845
$<X-B-X$	111.2°	110.4°	110.7°
$<C-P-C$	107.9°	107.2°	107.8°

reorganization of the BX_3 group is of importance in determining the acceptor strength of the boron halides (Rupp and Shriver, 1967; Brown, Drago and Bolles, 1968). CNDO/2 molecular orbital calculations indicated that BCl_3 has a significantly higher electron affinity than BF_3. It can thus be concluded that BCl_3 is a stronger electron acceptor than BF_3. The approximate molecular orbital calculations of Gropen and Haaland (1973) indicated that the amount of charge transferred from the donor to the acceptor was greater in $(CH_3)_3N.BCl_3$ than in $(CH_3)_3N.B(CH_3)_3$. It would be of interest to compare the values of $r(B-N)$ referring to vapour-phase $(CH_3)_3N.BF_3$ and $(CH_3)_3N.BCl_3$; however, the investigation of the latter has yet to be performed (cf., Table 9).

A comparison of the molecular geometries of the donor and acceptor parts of the vapour-phase addition compounds with those of their corresponding monomers can at present be made only for trimethylamine–borane (cf., Tables 5 and 10), and even then the planar configuration of the BH_3 molecule has to be assumed.* The deformation of the $H-B-H$ angle in the complex appears to be considerable. The bond angle of the donor, on the other hand, remains unchanged within the limit of experimental error. There is a significant lengthening of the $C-N$ bond.

In the light of the presently available structural information on the boron–nitrogen complexes, it is challenging to reconsider the puzzle posed by molecular geometry of trimethylamine–trimethylborane which generated much discussion almost 15 years ago (Lide, Taft and Love, 1959; Geller, 1960; Lide, 1960). Geller (1960) reproduced the

Ab initio calculations have recently been performed for BH_3 (Gelus and Kutzelnigg, 1973).

53

experimental value of the moment of inertia by a geometry with $r(B-N) = 1.62$ Å, in disagreement with the geometry advanced by Lide, Taft and Love (1959). Both sets of parameters are given below:

Assumed parameters	Lide *et al.* (1959)	Geller (1960)
$r(C-N)$	1.47 Å	1.50 Å
$r(C-B)$	1.56 Å	1.65 Å
$r(C-H)$	1.09 Å	1.09 Å
$<H-C-H$	108.5°	108.5°
$<C-N-C$	109.5°	105–106°
$<C-B-C$	109.5°	105–106°
Length of the B−N bond which, together with the assumed parameters, reproduces the experimental data	1.80 Å	1.62 Å.

The lengths of the C−N and C−B bonds are longer in the geometry of Geller (1960) than those found in the monomers. However, the value calculated for the moment of inertia was only weakly sensitive to these data (Lide, 1960). The assumption of considerably smaller than tetrahedral C−N−C and C−B−C bond angles was more important. Recent data for other compounds, particularly for $(CH_3)_3N.BH_3$ with a larger than tetrahedral H−B−H angle and nearly tetrahedral C−N−C angle (cf., Table 10), indicate that the model of Lide, Taft and Love (1959) is to be favoured.

BORON–PHOSPHORUS COMPLEXES

The number of volatile and stable boron–phosphorus compounds known to fulfill the prerequisites of the vapour-phase structural determinations regarding volatility and stability, is not very large. Nevertheless, the geometries of more compounds have already been elucidated in this group than among the boron–nitrogen complexes. More important, all of the studies have been performed by using modern techniques. The $r(B-P)$ bond distances, together with the data for the covalent and van der Waals linkages, are listed in Table 12.

TABLE 12

B−P bond lengths (Å) *in vapour-phase boron–phosphorus addition molecules*

Compound		$r(B-P)$	$\Sigma r_{cov.}$*	$\Sigma r_{v.d.W.}$*
$PF_3.BH_3$	(a1)	1.836 ± 0.012	1.98	3.4
$PHF_2.BH_3$	(a2)	1.832 ± 0.009		
$F_3P.BH_2BH_2.PF_3$	(b)	1.848 ± 0.028		
$(CH_3)_3P.BH_3$	(a3)	1.901 ± 0.007		
$CH_3PH_2.BH_3$	(a3)	1.906 ± 0.006		
$PH_3.BH_3$	(a4)	1.937 ± 0.005		

(a) Microwave spectroscopy.
   (1) Kuczkowski and Lide (1967).
   (2) Pasinski and Kuczkowski (1971).
   (3) Bryan and Kuczkowski (1972).
   (4) Durig, Li, Carreira and Odom (1973).
(b) Electron diffraction, Lory, Porter and Bauer (1971).
* $r_{cov.}$ = covalent radii, $r_{v.d.W.}$ = van der Waals radii, Pauling (1960).

Phosphorus trifluoride–borane, $F_3P.BH_3$, is a highly volatile but not very stable compound. Its microwave spectroscopic examination was carried out by Kuczkowski and Lide (1967) using both partially and fully deuterated samples. The structural data were obtained partly as r_s and partly as r_0 parameters. The validity of the assumptions concerning the internuclear distances, which remained unchanged upon isotopic substitution, was carefully examined and it was found that such an assumption can cause as much uncertainty as the experimental error. This conclusion was drawn from calculations assuming the average $r(B-H)$ value larger by 0.004 Å than the average $r(B-D)$ value. The structural parameters determined for $F_3P.BH_3$ are as follows:

$$
\begin{aligned}
r(B-H) &= 1.207 \pm 0.006 \text{ Å} \\
r(P-F) &= 1.538 \pm 0.008 \text{ Å} \\
r(B-P) &= 1.836 \pm 0.012 \text{ Å} \\
<D-B-D &= 115°4' \pm 1° \\
<F-P-F &= 99°50' \pm 1°
\end{aligned}
$$

No information could be obtained from the ground-state spectra as to whether the BH_3 and PF_3 groups are eclipsed or staggered. A potential barrier to internal rotation of 3240 ± 150 cal mol^{-1} was deduced from the vibrational satellite spectrum using the expression

55

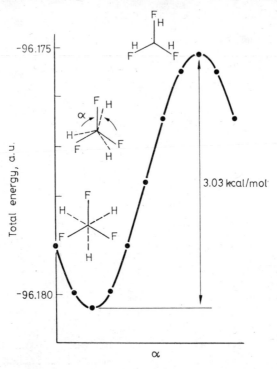

Fig. 8. The potential energy of the rotation around the B−P bond in $F_3P.BH_3$ as a function of the angle of rotation α calculated by the CNDO/2 method (Labarre and Leibovici, 1972)

$V = \dfrac{1}{2} V_0 (1 - \cos 3\alpha)$ for the potential function. In agreement with the microwave spectroscopic data, $V_0 = 3.03$ kcal mol^{-1} was obtained by Labarre and Leibovici (1972) on the basis of the CNDO/2 calculations. Part of the potential energy function versus the angle of rotation (α) around the B−P bond is shown in Fig. 8.

Difluorophosphine–borane, $HF_2P.BH_3$, was found to be more stable than $F_3P.BH_3$ (Rudolph and Parry, 1967). Six isotopic species were used in its structural determination by microwave spectroscopy (Pasinski and Kuczkowski, 1971). The most important findings can be summarized in the following way. The molecule has a staggered conformation with one plane of symmetry. The borane group is tilted away from the fluorine atoms, as indicated by the difference of approximately 10° between the two kinds of P–B–H bond angles:

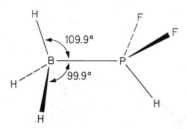

This difference is larger than would be expected as a consequence of experimental error or shrinkage effects. The phosphorus atom bond configuration is characterized by the following parameters:

$r(P-H) = 1.409 \pm 0.004$ Å $<B-P-H = 120.10 \pm 0.55°$
$r(P-F) = 1.552 \pm 0.006$ Å $<B-P-F = 117.73 \pm 0.28°$
$r(P-B) = 1.832 \pm 0.009$ Å $<F-P-F = 100.04 \pm 0.47°$
                                 $<F-P-H = 98.62 \pm 0.24°$

The range for the barrier to internal rotation was estimated to be 3600–4500 cal mol^{-1}, in agreement with the findings of later CNDO/2 calculations (Labarre and Leibovici, 1972).

Bis(trifluorophosphine)diborane(4), $F_3P.BH_2BH_2.PF_3$, is the only compound in this group whose molecular geometry was determined by electron diffraction (Lory, Porter and Bauer, 1971). The electron diffraction investigation was supplemented by gas-phase Raman and matrix isolation infrared spectroscopic studies. The latter was aimed primarily at distinguishing between the following three models:

(*i*) the two trifluorophosphine groups are in *syn* positions, the hydrogen atoms are terminal;

(*ii*) the two trifluorophosphine groups are in *anti* positions, the hydrogen atoms are terminal (model a);

(*iii*) the two trifluorophosphine groups are in *anti* positions, two of the hydrogen atoms are terminal and the other two are bridging (model b).

The first model was easily ruled out on the basis of comparisons between experimental and theoretical radial distributions and spectroscopic evidence. It was much more difficult to decide between the other two models (a) and (b):

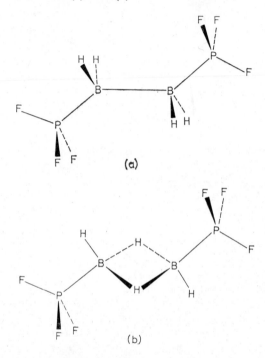

(a)

(b)

The least-squares method based on the molecular electron scattering intensities yielded a better agreement for model (a) than for model (b). The length of the $B-B$ bond was found to be 1.800 ± 0.012 Å. The following bond lengths and bond angles were obtained for both models; the uncertainties listed below are those of model (a), as favoured by the authors:

$$
\begin{aligned}
r(B-P) &= 1.848 \pm 0.009 \text{ Å} \\
r(P-F) &= 1.539 \pm 0.001 \text{ Å} \\
r(B-H), \text{ terminal} &= 1.252 \pm 0.054 \text{ Å} \\
<B-B-P &= 100.2 \pm 1.5° \\
<B-P-F &= 118.1 \pm 0.2°
\end{aligned}
$$

The spectra did not contain the band characterizing model (b) and consideration of physical and chemical evidence (Deever, Lory and Ritter, 1969) suggested also that model (a) was the better representation of the structure. It is interesting to note, in this connection, that the crystal-phase molecular structure of bis(triphenylphosphine)–diborane is consistent with model (a), although the authors of the X-ray diffraction study call for caution concerning the choice of that model (Vandoorne, Cordes and Hunt, 1973).

Trimethylphosphine–borane, $(CH_3)_3P.BH_3$, is formed by the reaction of $(CH_3)_3P$ and B_2H_6 and is stable enough for gas-phase structural determinations to be made. The other structural parameters were determined assuming values for the H . . . H non-bond distances of the methyl groups in the microwave spectroscopic analysis of Bryan and Kuczkowski (1972). The values obtained are as follows:

$$r(P-B) = 1.901 \pm 0.007 \text{ Å} \qquad <C-P-C = 105.0 \pm 0.4°$$
$$r(P-C) = 1.819 \pm 0.010 \text{ Å} \qquad <H-C-H = 109.3 \pm 1.0°$$
$$r(C-H) = 1.08 \ \ \pm 0.02 \ \ \text{Å} \qquad <H-B-H = 113.5 \pm 0.5°$$
$$r(B-H) = 1.212 \pm 0.010 \text{ Å}$$

Bryan and Kuczkowski (1972) reported on the structural analysis of *methylphosphine–borane*, $CH_3PH_2.BH_3$, in the same paper. One of the interesting questions posed by this molecule was whether a tilt of the BH_3 group similar to that in $HF_2P.BH_3$ would be observed. It was possible to determine an essentially r_s structure. The molecular conformation is characterized by the BH_3 and CH_3 groups staggering the P−H bonds:

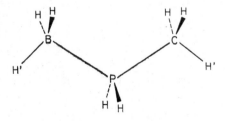

This finding agrees with that of the CNDO/2 calculation (Crasnier, Labarre and Leibovici, 1972). The most important bond lengths and bond angles were determined to be as follows:

$$r(P-C) = 1.809 \pm 0.006 \text{ Å}$$
$$r(P-B) = 1.906 \pm 0.006 \text{ Å}$$
$$r(P-H) = 1.404 \pm 0.006 \text{ Å}$$

$<H-P-H = 99.9 \pm 0.4°$	$<P-B-H' = 104.2 \pm 1.0°$
$<H-P-C = 103.2 \pm 0.6°$	$<P-C-H = 108.3 \pm 0.4°$
$<H-P-B = 116.3 \pm 0.6°$	$<P-C-H' = 111.3 \pm 1.0°$
$<P-B-H = 102.9 \pm 0.6°$	$<C-P-B = 115.7 \pm 0.4°.$

According to these data, there appears to be no pronounced tilt of the borane group in $CH_3PH_2.BH_3$ corresponding to that observed in $HF_2P.BH_3$. Taking the experimental errors into consideration, the difference of 1.3° between the two $P-B-H$ angles cannot be considered to be significant.

Phosphine–borane, $H_3P.BH_3$, is formed by the reaction of phosphine with diborane (McGandy, 1961; Rudolph, Parry and Farran, 1966). Durig, Li, Carreira and Odom (1973) determined its molecular geometry by microwave spectroscopy. The complete structure was obtained by analyzing the spectra of numerous isotopic species, and is characterized by the following parameters:

$$r(B-P) = 1.937 \pm 0.005 \text{ Å}$$
$$r(B-H) = 1.212 \pm 0.002 \text{ Å}$$
$$r(P-H) = 1.399 \pm 0.003 \text{ Å}$$

$<P-B-H = 103.6 \pm 0.2°$	$<H-B-H = 114.6 \pm 0.2°$
$<B-P-H = 116.9 \pm 0.2°$	$<H-P-H = 101.3 \pm 0.2°.$

The molecule has a staggered conformation and the barrier to internal rotation is 2.47 ± 0.05 kcal mol^{-1}.

The length of the $B-P$ bond is undoubtedly the most interesting parameter of the boron–phosphorus addition compounds. The gas-phase data in Table 12 are supplemented by the crystallographic results on $(CH_3)_3P.BX_3$ (X = Cl, Br, I) given in Table 11 and two more $r(B-P)$ values referring to the crystal phase presented here: 1.93 ± 0.01 Å in phosphine–borane (McGandy, 1961) and 1.887 ± 0.013 Å in $(NH_2)_3P.BH_3$ (Nordman, 1960). The variation of the $B-P$ bond distances in the $(CH_3)_3P.BX_3$ compounds shows a trend similar to that observed for the $(CH_3)_3N.BX_3$ complexes.

The data for $r(B-P)$ in $H_3P.BH_3$ indicate that the transition from the solid to the gaseous phase does not play an important role here. Our discussion will concentrate, however, on the vapour-phase data only.

One is easily tempted to use the experimental values of $r(B-P)$ for the formulation of various theories and hypotheses on bonding and also for testing them. Let us consider the compounds $F_3P.BH_3$ and $HF_2P.BH_3$ in order to illustrate this point. Some theories suggest that the $B-P$ bond of one of them is longer, while other approaches advocate just the opposite (for references see Pasinski and Kuczkowski, 1971). Within experimental error, however, the experimental $r(B-P)$ values are the same in these two molecules. A similar although more uncertain $r(B-P)$ value was found in $F_3P.BH_2BH_2.PF_3$. The $B-P$ bonds in $(CH_3)_3P.BH_3$ and $CH_3PH_2.BH_3$ can also be considered to be of the same length, while this bond is longer in the PH_3 derivative (Rudolph, Parry and Farran, 1966; cf. Table 12). Thus, two groups can be formed from the point of view of the $B-P$ bond length, even though the difference is not striking. Its trend, however, is in agreement with the prediction of various semi-empirical models. According to the valence shell electron pair repulsion (V.S.E.P.R.) theory, the presence of the strongly electronegative fluorine atom bonded to phosphorus causes a shortening of the $B-P$ bond. This theory, usually associated with Sidgwick and Powell (1938) and Gillespie and Nyholm (1957), was treated in detail in a book by Gillespie (1972). According to other considerations, the presence of the fluorine atom decreases the energy of the phosphorus d orbitals, making them more easily available for $d_\pi - p_\pi$ interactions, the effect appearing in the shortening of the $B-P$ bond.

That the longer donor–acceptor bond appears in the more stable $(CH_3)_3P.BH_3$ versus the shorter bond in the less stable $F_3P.BH_3$ is surprising. A detailed discussion of the issue was presented by Bryan and Kuczkowski (1972), who stressed that the usually accepted correlation between bond length and stability should be applied cautiously for addition compounds.

As already stated, it is also of interest to examine the structural changes of the donor and acceptor parts as compared with the monomers. Let us, then, examine the structural variations of PF_3 and $P(CH_3)_3$ upon complexation. As was seen, the changes in the $B-P$ bond distances follow the predictions of the V.S.E.P.R. model, and the same is true of the variations in the PF_3 and $P(CH_3)_3$ bond configurations. Morino, Kuchitsu and Moritani (1969), Moritani, Kuchitsu and Morino (1971) and Kuchitsu, Shibata, Yokozeki and Matsumura (1971) augmented the structural information on phosphorus halides with accurate data and showed convincingly the validity of the rules of the V.S.E.P.R. model for this group of compounds. The geometrical parameters for the fluorine derivatives, which previously were thought to require additional postulates (Gillespie, 1963), were also shown to fit the general V.S.E.P.R. pattern (Morino, Kuchitsu and Moritani, 1969).

The bond configurations of $R_3N.BH_3$ and $R_3P.BH_3$ (R = F or CH_3) are illustrated in Fig. 9, together with those of NR_3 and PR_3 and also of ONR_3 and OPR_3. If the electronegativity of the lone electron pair as "ligand" is considered to be zero, all of the variations in the bond lengths and bond angles are accounted for by the V.S.E.P.R. model in the case of the phosphorus compounds. On the other hand, the pattern for the nitrogen derivatives seems to be controversial. Although the comparison for such pairs of molecules as nitrogen trifluoride and trimethylamine or trifluoroamine oxide and trimethylamine oxide* fits perfectly into the V.S.E.P.R. theory, the same cannot be stated for the pair NF_3 and ONF_3 or $N(CH_3)_3$ and $ON(CH_3)_3$.

The geometry of trifluoroamine oxide indicates interesting bonding properties. According to the results of the electron diffraction study by Plato, Hartford and Hedberg (1970), the four ligands of the nitrogen atom are situated at the apexes of an almost regular tetrahedron. The molecular configuration has C_{3v} symmetry. The bond lengths indicate five covalent bonds formed by the nitrogen atom in ONF_3. Considering the molecular geometries of the phosphorus derivatives and ONF_3, it is expected that the $B-N$ bond in $F_3N.BH_3$, not yet studied, will be found to be considerably shorter than the $B-N$ bonds in the other boron–nitrogen complexes. No conclusion can be drawn from this prediction, however, concerning the stability of this presently hypothetical complex.

Comparing the structures of trimethylamine and trimethylamine–borane and also those of trimethylphosphine and trimethylphosphine–

---

* Trimethylamine oxide is the only compound in this series for which the structural determination refers to the crystalline state (Caron, Palenik, Goldish and Donohue, 1964). All other data originate from vapour-phase studies (for references, see Fig. 9).

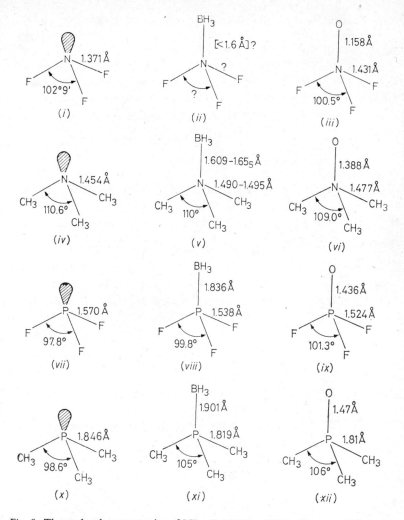

Fig. 9. The molecular geometries of NR_3, $R_3N.BH_3$, ONR_3, PR_3, $R_3P.BH_3$ and OPR_3 ($R = F$ or CH_3). The loop indicates a lone electron pair.

(*i*) Otake, Matsumura and Morino (1968); (*ii*) not yet studied; (*iii*) Plato, Hartford and Hedberg (1970); (*iv*) Beagley and Hewitt (1968); (*v*) Schirdewahn (1965); Durig, Li and Odom (1973); (*vi*) Caron, Palenik, Goldish and Donohue (1964); (*vii*) Morino, Kuchitsu and Moritani (1969); (*viii*) Kuczkowski and Lide (1967); (*ix*) Moritani, Kuchitsu and Morino (1971); (*x*) Bartell and Brockway (1960); (*xi*) Bryan and Kuczkowski (1972); (*xii*) Wang (1965)

63

borane, it can be concluded that the changes in these two pairs of compounds are of different character. The $C-N-C$ bond angles are the same in both $N(CH_3)_3$ and $(CH_3)_3N.BH_3$, whereas the $C-P-C$ bond angle is much larger in $(CH_3)_3P.BH_3$ than in $P(CH_3)_3$, as seen in Fig. 9. These results are in agreement with the rule of the V.S.E.P.R. theory according to which the repulsion of the electron pairs is stronger in a filled valence shell than in an incompletely filled shell. Accordingly, the angles formed by the bonds of the nitrogen atom (second period) will not deviate much from the tetrahedral value, while those of phosphorus (third period) will change more readily. However, no such simple interpretation can be given to the observation that the $C-N$ bond is longer in $(CH_3)_3N.BH_3$ than in $N(CH_3)_3$, whereas the $C-P$ bond is shorter in $(CH_3)_3P.BH_3$ than in $P(CH_3)_3$.

ALUMINIUM–NITROGEN COMPLEXES

The first reports of studies on the vapour-phase molecular geometries of aluminium–nitrogen complexes appeared in 1972 and the structures of only the following four compounds have been published:

Aluminium trichloride–ammonia, $Cl_3Al.NH_3$
Aluminium trichloride–trimethylamine, $Cl_3Al.N(CH_3)_3$
Alane–trimethylamine, $H_3Al.N(CH_3)_3$
Trimethylaluminium–trimethylamine, $(CH_3)_3Al.N(CH_3)_3$

(for references see Table 13). The results of the electron diffraction investigations are summarized in Table 13.

The following assumptions were typical in the structural analyses of the $X_3Al.NR_3$ compounds:

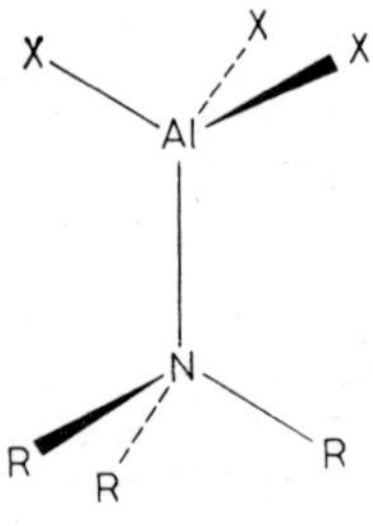

TABLE 13

Vapour-phase molecular geometries* of aluminium–nitrogen addition compounds as determined by electron diffraction

$AlX_3.NR_3$	$AlCl_3.NH_3$ (a)	$AlCl_3.N(CH_3)_3$ (b)	$AlH_3.N(CH_3)_3$ (c)	$Al(CH_3)_3.N(CH_3)_3$ (d)
$r(Al-N)$, Å	1.996 ± 0.019	1.945 ± 0.035	2.063 ± 0.008	2.099 ± 0.010
$r(Al-X)$, Å	2.100 ± 0.005	2.121 ± 0.004	1.560 ± 0.011	1.987 ± 0.005
$<X-Al-X$	$116.3_5 \pm 0.4°$	$(113.6°)**$	$(114.1°)**$	$(115.6°)**$
$<X-Al-N$	$(101.0°)**$	$104.9 \pm 0.7°$	$104.3 \pm 1.1°$	$102.3 \pm 0.3°$
$r(C-N)$, Å		1.516 ± 0.012	1.476 ± 0.003	1.474 ± 0.003
$<C-N-Al$		$112.6 \pm 1.5°$	$109.0 \pm 0.3°$	$109.3 \pm 0.4°$

* r_a parameters.
** Calculated from the other angle by virtue of C_{3v} symmetry.
(a) Hargittai, Hargittai and Spiridonov (1973).
(b) Almenningen, Haaland, Haugen and Novak (1973).
(c) Almenningen, Gundersen, Haugen and Haaland (1972).
(d) Anderson, Forgaard and Haaland (1972).

(*i*) the molecules have C_{3v} symmetry as a whole,

(*ii*) all of the C–H bonds in the molecule are of the same length,

(*iii*) the methyl groups have C_{3v} local symmetry and the symmetry axis coincides with the Al–C or N–C bonds,

(*iv*) the angle of rotation around the Al–C or N–C bonds is such that the C–H bonds are staggered with respect to the Al–N bond.

It was also typical to perform a least-squares analysis on the parameters for a model in which the donor ligands staggered those of the acceptor. This choice for molecular conformation was based partly on structural information on analogous molecules and partly on the results of the structural analyses themselves, as the models with staggered forms showed better agreement with the experimental data than those with eclipsed forms. Thus, *e.g.*, the eclipsed conformation for aluminium trichloride–trimethylamine could be ruled out on the basis of comparisons between experimental and theoretical radial distributions (Almenningen, Haaland, Haugen and Novak, 1973). It is much more difficult to gain information on the molecular conformation of these complexes in the presence of hydrogen atoms as ligands, as the atomic pairs that contain the hydrogen atom have a relatively small influence in electron scattering. As the associated mean amplitudes of vibration are particularly large for the distances that involve hydrogen atoms, the conformational analysis is even more difficult. The eclipsed form of alane–trimethylamine, for instance, was found to approximate the experimental data better than the staggered form in the study of Almenningen, Gundersen, Haugen and Haaland (1972). The $N-Al-H$ bond angle, however, became unrealistically large and the model was rejected. Aluminium trichloride–ammonia has a staggered conformation and the barrier to the rotation from the staggered into the eclipsed form is 2.11 kcal mol^{-1} according to CNDO/2 calculations (Hargittai, Hargittai, Spiridonov, Pelissier and Labarre, 1975).

To supplement the data on the $r(Al-N)$ values in Table 13 referring to the vapour phase, the results of two crystal-phase X-ray diffraction determinations are given here:

$$Cl_3Al.N(CH_3)_3, \quad r(Al-N) = 1.96 \pm 0.01 \text{ Å}$$
(Grant, Killean and Lawrence, 1969);

$$(BH_4)_3Al.N(CH_3)_3, \quad r(Al-N) = 1.99 \pm 0.01 \text{ Å}$$
(Bailey, Bird and Wallbridge, 1968).

As expected, the $Al-N$ bond shortens as the electronegativity of the ligand (X) of aluminium increases. On the other hand, the increase in electronegativity of the ligand (R) of nitrogen is accompanied by lengthening of the $Al-N$ bond. The two phenomena are connected with the increasing acceptor character of the X_3Al part and the decreasing donor character of the NR_3 part, as pointed out by Almenningen, Gundersen, Haugen and Haaland (1972). A similar observation was made by Stucky (1974). The geometry of the other parts of the complex molecule is also influenced, of course, by the strength of the acceptor and donor parts. Thus, $e.g.$, the $Al-Cl$ bond is longer in $Cl_3Al.N(CH_3)_3$ than in $Cl_3Al.NH_3$ or in crystalline $Cl_3Al.CH_3CH_2COCl$ as a consequence of the increased donor character. The value of $r(Al-Cl)$ was determined to be 2.094 Å for the latter compound (Le Carpentier and Weiss, 1972). Similarly, as the acceptor character increases, the $C-N$ bond appears to be longer in $Cl_3Al.N(CH_3)_3$ than in $(CH_3)_3Al.N(CH_3)_3$ (cf., Table 13).

A gradual lengthening of the bonds between the atoms of four-coordinated aluminium and also four-coordinated nitrogen was observed by McDonald and McDonald (1972), in the series of AlN,

RAlNR, R_2AlNR_2 and R_3AlNR_3, as illustrated by the following data:

aluminium–nitride, $r(Al-N) = 1.893$ Å
(Jeffrey, Parry and Mozzi, 1956);
$[C_6H_5AlNC_6H_5]_4$, $r(Al-N) = 1.914$ Å
(McDonald and McDonald, 1972);
$[(CH_3)_2AlNHCH_3]_2$, $r(Al-N) = 1.953$ Å
(Gosling, McLaughlin, Sim and Smith, 1970);
$(CH_3)_3AlN(CH_3)_3$, $r(Al-N) = 2.10$ Å
(Anderson, Forgaard and Haaland, 1972).

The first three compounds above have been studied in the crystal phase by X-ray diffraction.

The results of crystal-phase X-ray diffraction studies (Semenenko, Lobkovskii and Dorosinskii, 1972) and vapour-phase electron diffraction studies (Semenenko, Lobkovskii, Bulichev, Golubinskii, Vilkov and Mastryukov, 1973) on the molecular structure of

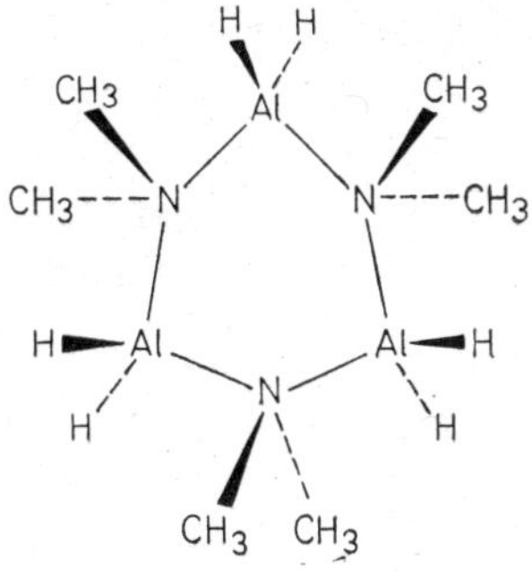

are still a matter of controversy. The values obtained for $r(Al-N)$ fit the above data well (1.93 Å and 1.97 Å from the two studies in the order mentioned above). According to CNDO/2 calculations, a strong $Al \cdots Al$ $\cdots Al \cdots$ transangular bonding interaction may play an important role in stabilizing a preferred conformation in this compound (Pelissier, Labarre, Vilkov, Golubinsky and Mastryukov, 1974).

It is of interest to compare the crystal- and vapour-phase data for $Cl_3Al.N(CH_3)_3$. There is good agreement on the shape and most of the geometrical parameters of the molecule. For example, the mean value for $r(Al-Cl)$ obtained for the crystal, 2.123 Å (Grant, Killean and Lawrence, 1969), is the same as the electron diffraction result for this parameter (Table 13). Certain ambiguity exists, however, for detailed comparison when the differences between the crystallographically non-equivalent parameters are larger than the deviation of their mean value from the vapour-phase value (see, *e.g.*, the $C-N-C$ bond angles).

Concerning the donor and acceptor geometries as compared with the structures of the corresponding uncomplexed molecules, again small deformations for the donor parts and larger changes for the acceptor parts are observed (cf., Tables 5, 7 and 13). The following general explanation is suggested in order to account for this phenomenon. The lone electron pair of the donor is attracted by the acceptor as the complexation is taking place and, consequently, the repulsion from the lone pair towards the bonding pairs of the donor is decreasing. This effect tends to increase the bond angles between the donor ligands. On the other hand, the coordination linkage formed is repelling the bonding pairs of the acceptor, resulting in a decrease in the bond angles between the acceptor ligands. Another consequence of the complexation is the appearance of new non-bond interactions that tend to decrease the bond angles in both the donor and acceptor parts. Hence, all of the changes due to the complexation have the effect of closing the acceptor bond angles and the geometry becomes markedly pyramidal from planar or nearly planar. It is not so easy to predict the variation in the donor bond configuration because it appears as a result determined by the two competing effects described. This is illustrated in Fig. 10.

The comparison of the molecular geometries of free and complex molecules also has importance in the discussion on the structure of monomeric $AlCl_3$ (cf., p. 46). The bond angle in $AlCl_3$ cannot be expected to be smaller than the $Cl-Al-Cl$ bond angle in the addition complexes of aluminium trichloride. As the controversy centred around the equilibrium or, at least, average structure of $AlCl_3$, it is of interest to have the internuclear distances in terms of the r_α structure for the aluminium bond configuration of $Cl_3Al.NH_3$. This was obtained by using the $K = [\langle(\Delta x)^2\rangle + \langle(\Delta y)^2\rangle]/2r$ values from Cyvin, Cyvin and Hargittai (1974) and ignoring the δr terms in the expression $r_\alpha \approx r_a + \dfrac{l^2}{r} - \delta r - K$. The following values were obtained (T = 522 K,

68

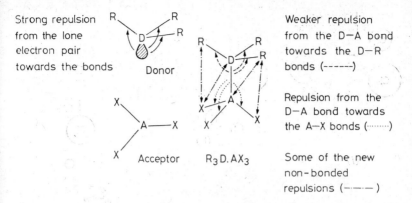

Strong repulsion from the lone electron pair towards the bonds — Donor

Weaker repulsion from the D–A bond towards the D–R bonds (------)

Repulsion from the D–A bond towards the A–X bonds (·········)

Some of the new non-bonded repulsions (—·—·—)

Acceptor $R_3D.AX_3$

Fig. 10. Illustration for the discussion on the geometrical consequences of the formation of charge-transfer complexes

the nozzle temperature of the electron diffraction experiment):

Atomic pair	r_a (Å)	K (Å)	r_α (Å)
Al – Cl	2.100	0.006	2.096
Cl . . . Cl	3.569	0.002	3.572
Al – N	1.996	0.004	1.994
Cl . . . N	3.165	0.005	3.173

The corresponding average angle for Cl–Al–Cl is 116.9°. This flat pyramidal average configuration of the Cl_3Al part of the complex also suggests a planar average structure or little deviation from planarity for free $AlCl_3$ (Hargittai, Hargittai, Spiridonov, Pelissier and Labarre, 1975).

In addition to the geometrical variations, it is of interest to investigate the effects of complexation on the molecular force fields and vibrations. Such studies have been initiated for some simple models (Cyvin and Andreassen, 1974), including $Cl_3Al.NH_3$ (Hargittai, Hargittai, Cyvin and Cyvin, 1974; Cyvin, Cyvin and Hargittai, 1974).

Returning to the discussion of the geometrical variations, the bond configuration of the donor (acceptor) part in the charge-transfer complex is expected to be intermediate between those of the monomeric molecule and the analogous positive (negative) ion, as pointed out by Almenningen, Gundersen, Haugen and Haaland (1972).

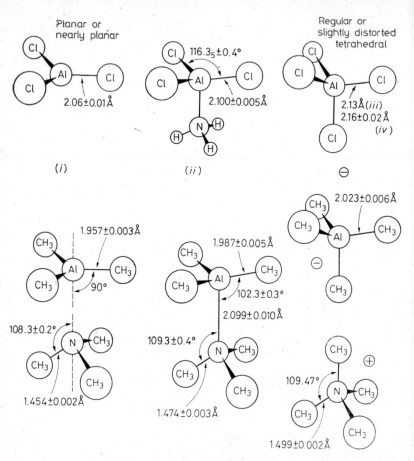

Fig. 11. Comparison of the molecular geometries of isolated molecules, ions and charge-transfer complexes.
(*i*) Zasorin and Rambidi (1967); (*ii*) Hargittai, Hargittai and Spiridonov (1973); (*iii*) Stork-Blaisse and Romers (1971); (*iv*) Spiridonov, Erokhin and Lutoshkin (1971).
For references not indicated see in Anderson, Forgaard and Haaland (1972) and in the text

Figure 11 demonstrates these relationships after Anderson, Forgaard and Haaland (1972) and Hargittai, Hargittai and Spiridonov (1973) for some of the aluminium–nitrogen complexes.

The degree of the charge-transfer itself is also of interest in the addition compounds and was calculated to be 0.38, 0.19 and 0.47 for $Cl_3Al.NH_3$ (Hargittai, Hargittai, Spiridonov, Pelissier and Labarre, 1975), $PH_3.BH_3$ and $PHF_2.BH_3$ (Labarre and Leibovici, 1972) and $NH_3.BH_3$ (Bach, Crasnier, Labarre and Leibovici, 1973), respectively.

MISCELLANEOUS

Durig, Hudgens and Odom (1974) carried out detailed spectroscopic studies on the structure of trimethylarsine–borane, $(CH_3)_3As.BH_3$. The vibrational spectra of low-temperature solid-state samples were interpreted on the basis of a model with C_{3v} symmetry. The microwave spectrum of four isotopic species was analyzed. Assuming $r(C-H) = 1.09$ Å, $H-C-H$ bond angle = 107°, $r(B-H) = 1.212$ Å and $H-B-H$ bond angle = 113.5° (for the latter two, see the results on $(CH_3)_3P.BH_3$, p. 59) the following parameters were determined:

$$\begin{aligned}
r(As-B) &= 2.035 \text{ Å} \\
r(As-C) &= 1.945 \text{ Å} \\
<C-As-C &= 105°.
\end{aligned}$$

The arsine–boron linkage was observed to be only slightly longer than the sum of the covalent radii (2.02 Å). The $C-As-C$ angle opens considerably and the $As-C$ shortens a little compared with the geometry of the uncomplexed trimethylarsine, $As(CH_3)_3$, $C-As-C$ bond angle = $96 \pm 3°$ and $r(As-C) = 1.959 \pm 0.01$ Å (Lide, 1959).

Spiridonov and Malkova (1969) studied the vapour-phase molecular geometry of *aluminium tribromide–antimony tribromide* by electron diffraction. Two molecular models were examined in detail:

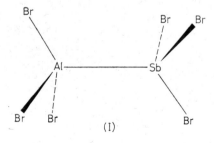

(I)

71

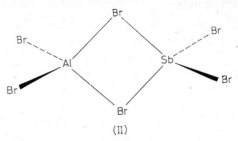

(II)

Good agreement between the experimental and theoretical distributions was obtained for model (I). The bond configuration of the $SbBr_3$ part was found to be essentially the same as in the isolated $SbBr_3$ molecule (Konaka and Kimura, 1973). The geometry of the $AlBr_3$ part (Br$-$Al$-$Br bond angle = 110 ± 2°), on the other hand, differed considerably from the planarity generally accepted for the isolated $AlBr_3$ molecule. The aluminium bond lengths were found to be $r(Al-Sb) = 2.52 \pm 0.02$ Å and $r(Al-Br) = 2.30 \pm 0.02$ Å. The latter is intermediate between the corresponding parameters determined in dimeric aluminium tribromide, Al_2Br_6, for the bridging and terminal bonds (Akishin, Rambidi and Zasorin, 1959).

As the V. S. E. P. R. model is applied to the molecular geometry of this compound, the following model can be postulated:

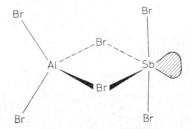

assigning a trigonal bipyramidal configuration for the antimony bonds with the lone pair of electrons in an equatorial position. This model is to be tested against new experimental data (Spiridonov, 1974).

The gallium bond configuration in *gallium trichloride–ammonia*, $Cl_3Ga.NH_3$, was elucidated by Hargittai, Hargittai and Spiridonov (1976) in a study similar to that for $Cl_3Al.NH_3$. The internuclear distances (r_a parameters) were found to be as follows:

$$r(Ga-Cl) = 2.142 \pm 0.005 \text{ Å}$$
$$r(Cl \ldots Cl) = 3.642 \pm 0.010 \text{ Å}$$

$$r(\text{Ga} - \text{N}) = 2.057 \pm 0.011 \text{ Å}$$
$$r(\text{Cl} \ldots \text{N}) = 3.242 \pm 0.012 \text{ Å}.$$

The corresponding value for Cl – Ga – Cl is 116.4°. As no spectroscopic analysis has been performed on this compound yet, the spectroscopic correction terms are not available.

The difference between the values of $r(\text{Ga} - \text{Cl})$ and $r(\text{Al} - \text{Cl})$ in the complexes ($\Delta r = 0.042$ Å) is only slightly greater than the analogous difference between the lengths of the terminal bonds in Ga_2Cl_6 and Al_2Cl_6 (see Tables 24 and 25), i.e., 0.034 Å. The value of $\Delta r = r(\text{Ga} - \text{N}) - r(\text{Al} - \text{N})$ is, however, 0.06 Å. Although the uncertainties for the latter bond distances are large, the appearance that the gallium–nitrogen linkage is relatively weaker than the aluminium–nitrogen linkage seems to be a real effect. This observation, based on geometrical evidence, is convincingly supported by the results of mass spectrometric studies, which indicated that the Ga – N bond is relatively weaker with respect to the Ga – Cl bond in $\text{Cl}_3\text{Ga.NH}_3$ than is the Al – N bond with respect to the Al – Cl bond in $\text{Cl}_3\text{Al.NH}_3$ (Hargittai, Hargittai, Tamás, Bihari, Spiridonov and Ivanov, 1974).

We conclude this section by mentioning the X-ray diffraction studies on the crystal-phase molecular structures of $\text{Cl}_3\text{As.N(CH}_3)_3$ (Webster and Keats, 1971) and $\text{Cl}_4\text{Ge.N(CH}_3)_3$ (Bilton and Webster, 1972). The trigonal bipyramidal configurations (Fig. 12) observed in both instances are in complete agreement with the V.S.E.P.R. model.

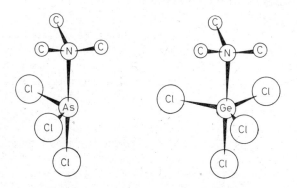

Fig. 12. The molecular models (without the hydrogen atoms) of $\text{Cl}_3\text{As.N(CH}_3)_3$ (Webster and Keats, 1971) and $\text{Cl}_4\text{Ge.N(CH}_3)_3$ (Bilton and Webster, 1972)

73

ELECTRON-DEFICIENT MOLECULES

The structures to be discussed in this chapter are usually called electron-deficient in that there are too few valence electrons to permit two electrons per bond. The concept of multi-centre bonds is used to interpret this phenomenon. In the three-centre bonds, for example, three atomic orbitals overlap to form one bonding orbital with two electrons and two anti-bonding orbitals with no electrons.

The electronic and geometrical structures of electron-deficient molecules have posed a challenge to structural chemists for many years. Detailed knowledge has already been accumulated on the structures of certain classes of compounds, *e.g.*, boron hydrides, based mainly on crystal-phase X-ray diffraction studies (see, *e.g.*, Lipscomb, 1963). For other classes of compounds, such as metal borohydrides, the discussions on controversial issues to be found in the literature indicate the need for further painstaking studies.

The results of vapour-phase investigations on boron hydrides, carboranes, metal borohydrides and some metal–alkyl bridge structures are reviewed in this chapter. Microwave spectroscopy has played an important role in elucidating many vapour-phase molecular geometries among the compounds dealt with here. Some of the general merits and limitations of microwave spectroscopy have already been mentioned. There are two particular advantages that are worth noting in connection with the boron hydrides and carboranes. One is the natural occurrence of boron as two isotopes, ^{10}B and ^{11}B, with relative abundances of 19% and 81%, respectively. Accordingly, a large variety of isotopic species exists in sufficient quantity to be spectroscopically visible when the vapours of boron hydrides and carboranes are studied by microwave spectroscopy. The relative stiffness of the cage-like structures provides the other advantage, as the geometrical changes in isotopically substituted species due to the zero-point vibrations will not be considerable.

74

There are still various difficulties from the point of view of the application of physical techniques, which explain the scarcity of vapour-phase structural data for boron hydrides. These difficulties include the lack of a permanent electric dipole moment in some compounds, thus preventing microwave spectroscopic studies, and the relatively small scattering ability of the hydrogen atom, which makes electron diffraction structural determinations more uncertain. The increasing complexity of these molecules is another cause of difficulty, although some of the rather complicated carborane derivatives have been investigated. The small carboranes have been studied almost exclusively by microwave spectroscopy, while electron diffraction has been applied to a series of larger carborane molecules.

The structural elucidation of the metal borohydrides is not an easy task because often even the molecular shape is unknown. Only very few compounds have been studied so far, some of which, however, underwent repeated investigations in order to obtain more accurate determinations of fine detail or to decide controversial interpretations. The structural studies on electron-deficient compounds often provide examples of successfully combined applications of various physical techniques. Here, only the investigation of small carboranes by magnetic resonance spectroscopy and microwave spectroscopy (Onak, Dunks, Beaudet and Poynter, 1966) and the studies on metal borohydrides involving electron diffraction and vibrational spectroscopy are mentioned.

BORON HYDRIDES

The simplest boron hydride, BH_3, was observed as a molecule by mass spectrometry for the first time in 1964 (Sinke, Pressley, Baylis and Stafford, 1964; Fehler and Koski, 1964). It is generally assumed that this molecule has a trigonal planar configuration (D_{3h}).

The hydrogen-bridge structure (Fig. 13) of *diborane*, B_2H_6, is firmly established and well known. It is one of the simplest and probably the most thoroughly studied of the electron-deficient molecules. Its molecular geometry, as is now generally accepted, was finally elucidated by vibrational spectroscopic analysis by Price (1947, 1948) and electron diffraction determinations by Hedberg and Schomaker (1951) after much discussion and numerous reinvestigations (see the references in Hedberg and Schomaker, 1951, and Bartell and Carroll,

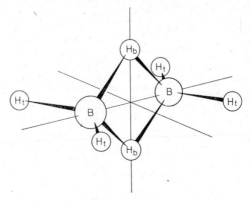

Fig. 13. The molecular configuration of diborane, B_2H_6

1965a). Hedberg and Schomaker (1951) used the visual technique of electron diffraction. The results of their early study, however, successfully passed the test of the more accurate sector-microphotometer studies (Bartell and Carroll, 1965a). Satisfactory as they may be, even with up-to-date standards, the results of the visual electron diffraction studies concern only the molecular shape and internuclear distances. No information can be obtained on the intramolecular motion from the visual work. The study by Bartell and Carroll (1965a) was partly aimed, accordingly, at gaining information on the intramolecular motion of diborane. Their other purpose was to determine the effects of isotopic substitution on molecular structure by simultaneously analyzing the data recorded for heavy diborane, B_2D_6. Only partial results were obtained in this respect, however, because experimental errors prevented the desired accuracy being achieved.

The modern work of Bartell and Carroll (1965a) provided geometrical parameters with well defined physical significance. The electron diffraction results were characterized by r_g parameters that are the thermal average values of the internuclear distances. Using these results,*

---

* As there was a strong correlation between the difference of the bridge and terminal B—H distances and the mean amplitudes of vibration of these bonds (Bartell and Carroll, 1965a), and, as the calculated amplitudes (Kuchitsu, 1968) were slightly but consistently larger than the experimental values, corrections were applied to the electron diffraction r_g values by Kuchitsu (1968) before using them in the determination of the r_z structure.

TABLE 14

Geometrical parameters (a) for the B_2H_6 molecule as determined by electron diffraction and infrared spectroscopy

	r_g (b)	r_0 (c)	r_z (d)
$r(B-H_t)$	1.196_0 (+0.008, −0.006) Å	$1.200_5 \pm 0.0018$ Å	$1.19_2 \pm 0.01$ Å
$r(B-H_b)$	1.339_3 (+0.002, −0.006) Å	$1.320_4 \pm 0.0005$ Å	$1.32_9 \pm 0.005$ Å
$r(B-B)$	$1.775_0 \pm 0.003_5$ Å	$1.762_8 \pm 0.0013$ Å	$1.77_0 \pm 0.005$ Å
$r(B...H_t)$	$2.588_1 \pm 0.009$ Å		
$<H_t-B-H_t$		$121.0 \pm 0.3°$	$121._8 \pm 3°$
$<H_b-B-H_b$		$96.2 \pm 0.1°$	$96._5 \pm 0.5°$

(a) Here and henceforth, H_t = terminal hydrogen atom, H_b = bridge hydrogen atom.
(b) Electron diffraction, Bartell and Carroll (1965a).
(c) Effective parameters obtained from the infrared spectroscopic investigation, Coyle, Lafferty and Maki (1967).
(d) Results from combining electron diffraction and spectroscopic data, Kuchitsu (1968).

TABLE 15

r_g internuclear distance parameters (Å) for the B_2D_6 molecule determined by electron diffraction, Bartell and Carroll (1965a)

$r(B-D_t)$	1.198_0 (+0.006, −0.005)
$r(B-D_b)$	1.333_5 (+0.002, −0.004)
$r(B-B)$	$1.771_2 \pm 0.003_5$
$r(B...D_t)$	$2.572_3 \pm 0.008$

together with the r_0 effective structure (for the $B-B$ bond, an r_s parameter) from the high-resolution infrared spectroscopic study of Coyle, Lafferty and Maki (1967), the average structure for the ground vibrational state (r_z) was determined by Kuchitsu (1968). The geometrical parameters obtained for B_2H_6 are summarized in Table 14. The internuclear distances of B_2D_6, as determined by electron diffraction, are given in Table 15.

The mean amplitudes of vibration determined for diborane (Bartell and Carroll, 1965a) are worthy of more attention. The electron diffraction radial distributions are reproduced in Fig. 14. It can easily be seen that the peaks assigned to the internuclear distances involving the

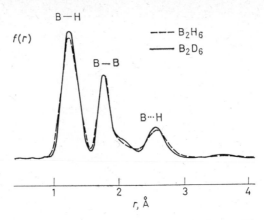

Fig. 14. The radial distributions obtained from the electron diffraction data on B_2H_6 and B_2D_6 (Bartell and Carroll, 1965a)

hydrogen atom are lower and broader for B_2H_6 than the corresponding peaks for B_2D_6. As pointed out by Bartell and Carroll (1965a), this effect is a consequence of the zero-point vibrations. The hydrogen atoms are lighter than deuterium atoms. The corresponding vibrational frequencies and zero-point energies are thus higher, and the amplitudes of vibration larger.

The intramolecular motion of diborane is characterized by the vibrational amplitudes of the bridge bonds (l_b) being larger than those of the terminal bonds (l_t). The longer bridge bonds are thus looser than the shorter terminal bonds. The amplitude ratio was found to be the same for B_2H_6 and B_2D_6: $l_b/l_t = 1.17 \pm 0.06$ and 1.18 ± 0.05, respectively. Bartell and Carroll (1965a) suggested rationalizing the magnitude in terms of Badger's rule (Badger, 1934), establishing the relationship

$$k = 1.86 \times 10^5 \, (r - d_{ij})^{-3}$$

between the force constant k and the bond length r with $d_{ij} = 0.34$ Å for the boron–hydrogen bonds. Using the expression relating the zero-point amplitude to the force constant:

$$l = (16\pi^2 \mu k/h^2)^{-\frac{1}{4}}$$

78

which is valid for diatomic molecules with a reduced mass of μ, the equation

$$\frac{l_b}{l_t} = \left(\frac{r_b - 0.34}{r_t - 0.34} \right)^{\frac{3}{4}}$$

was obtained, yielding the value of 1.12 ± 0.01 for both B_2H_6 and B_2D_6. The agreement between the values obtained from the experimental data and those estimated from the above calculation provided additional support to the reality of the difference between the l_b and l_t values determined for diborane. Cyvin and Vizi (1969) calculated somewhat smaller values for the ratio l_b/l_t: 1.03 and 1.04 for B_2H_6 and B_2D_6, respectively. The determined and calculated amplitudes of vibration from two independent studies are presented in Table 16.

TABLE 16

Mean amplitudes of vibration (Å) of the B_2H_6 and B_2D_6 molecules determined by electron diffraction (ED) and calculated from spectroscopic (SP) data

	ED	SP			
	(a)	0 K (b)	298 K (b)	0 K (c)	298 K (c)
B_2H_6					
$B-H_t$	$0.073_4 \pm 0.006$	0.084_5	0.084_5	0.0919	0.0920
$B-H_b$	$0.085_7 \pm 0.009$	0.101_5	0.101_5	0.0950	0.0951
$B-B$	$0.060_8 \pm 0.002$	0.059_9	0.061_1	0.0565	0.0573
$B \ldots H_t$	$0.117_8 \pm 0.005$	0.129_1	0.122_7	0.1264	0.1332
B_2D_6					
$B-D_t$	$0.064_3 \pm 0.005$	0.072_3	0.072_3	0.0786	0.0787
$B-D_b$	$0.075_6 \pm 0.006$	0.086_8	0.087_1	0.0815	0.0818
$B-B$	$0.059_5 \pm 0.002$	0.059_4	0.060_8	0.0560	0.0570
$B \ldots D_t$	$0.104_9 \pm 0.005$	0.106_4	0.117_7	0.1091	0.1202

(a) The temperature of the electron diffraction experiment was 25°C, Bartell and Carroll (1965a).
(b) Kuchitsu (1968).
(c) Cyvin and Vizi (1969).

In conclusion, we mention the recent review paper on diborane by Long (1972), which also summarizes the calculations and the theoretical studies.

The first structural elucidation of *bromodiborane*, B_2H_5Br,

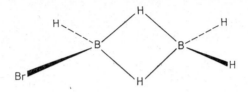

was performed by using microwave spectroscopy by Cornwell (1950). The up-to-date version of the same technique was used by Ferguson and Cornwell (1970) for a more detailed analysis 20 years later. One of the interesting features of this study was the two independent determinations of $r(B-B)$, $r(B-Br)$ and the $B-B-Br$ bond angle in terms of the r_s structure using the isotopic substitution method. The two sets of parameters are identical within the estimated error, as demonstrated by the following data:

	Values from $^{11}B_2H_5^{81}Br$-based data	Values from $^{10}B_2H_5^{81}Br$-based data	Average values
$r_s(B-B)$, Å	$1.773_8 \pm 0.002_8$	$1.771_3 \pm 0.002_7$	1.773 ± 0.003
$r_s(B-Br)$, Å	$1.927_9 \pm 0.006_3$	$1.931_5 \pm 0.004_9$	1.930 ± 0.005
$<B-B-Br$	$121.47 \pm 0.33°$	$121.30 \pm 0.25°$	$121.4 \pm 0.3°$

The distance between the two bridge hydrogen atoms was also obtained (1.954 ± 0.002 Å) in addition to the parameters listed. The other parameters could not be determined, however, owing to a lack of more independent data. It was only established that the electron diffraction results for $r(B-H_t)$ and the $B-B-H_t$ bond angle of diborane (Bartell and Carroll, 1965a) were consistent with the experimental values of moments of inertia obtained for bromodiborane.

Ferguson and Cornwell (1970) estimated the corrections for converting the $r_s(B-B)$ and $r_0(H_b \ldots H_b)$ parameters into r_z data. These estimations were based on the comparison of the molecular geometries of bromodiborane and diborane. This analysis was aimed at producing a comparison of the structures obtained by electron diffraction for B_2H_6 and by microwave spectroscopy for B_2H_5Br. The most realistic means of comparison is in terms of the r_z structure for which the effects of molecular vibrations are minimized. The r_z parameters of

80

diborane are shown in Table 14, and the following corrected values were obtained for B_2H_5Br:

$$r_z(B-B) = 1.773 \pm 0.015 \text{ Å}$$
$$r_z(H_b \ldots H_b) = 1.99_2 \pm 0.02_6 \text{ Å}.$$

As shown by the r_z data, the $B-B$ and $H_b \ldots H_b$ internuclear distances change negligibly upon substitution of bromine for terminal hydrogen. Similarly, the variations in the $C-C$ bond distances in ethylene and its derivatives are small (Ferguson and Cornwell, 1970). The available microwave spectroscopic data may still allow the interpretation that the bridge hydrogen atoms in B_2H_5Br are not equidistant from the boron atoms, or that the bridge hydrogen atoms and the boron atoms are not coplanar.

The molecular geometry of *tetramethyldiborane*, $(CH_3)_2BH_2B(CH_3)_2$, was determined by the electron diffraction study of Carroll and Bartell (1968). The molecular shape is analogous to that of diborane:

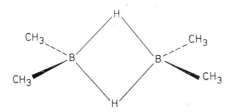

The value $r_g(B-C) = 1.590 \pm 0.003$ Å is larger than that found in $B(CH_3)_3$ (see Table 4). The $B-B$ bond in tetramethyldiborane $(1.840 \pm 0.010$ Å) is considerably longer than the analogous bond in diborane (see Table 14). The applicability of different considerations on bonding were discussed by Carroll and Bartell (1968) in their comparison of the structures of diborane and its derivatives with those of hydrocarbons.

Using the visual technique of electron diffraction, Hedberg and Stosick (1952) verified that the substituent occupies the position of one of the two bridge hydrogen atoms in *aminodiborane*, $H_2NB_2H_5$, and *N,N-dimethylaminodiborane*, $(CH_3)_2NB_2H_5$. The molecular configuration of $(CH_3)_2NB_2H_5$ is shown in Fig. 15. Assuming C_{2v} sym-

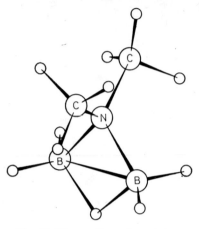

Fig. 15. The configuration of the
$(CH_3)_2NB_2H_5$ molecule

metry for the molecular model of $H_2NB_2H_5$ and the values $r(N-H) =$ $= 1.02$ Å, $r(B-H_b) = 1.35$ Å, $H-N-H$ bond angle $= 109.5°$ and $H-B-H$ bond angle $= 120°$, the following bond lengths and bond angle were determined:

$$
\begin{aligned}
r(B-N) &= 1.50_4 \pm 0.02_6 \text{ Å} \\
r(B-B) &= 1.93 \pm 0.09 \text{ Å} \\
r(B-H_t) &= 1.15 \pm 0.09 \text{ Å} \\
<B-N-B &= 76.2 \pm 2.8°.
\end{aligned}
$$

The angle ε between the extension of the $B-B$ bond and the plane of the BH_2 group, defined to be increasing as BH_2 rotates away from N, was found to be $15 \pm 20°$.

A reinvestigation of the dimethyl derivative was performed by microwave spectroscopy by Cohen and Beaudet (1973). Their results, together with those of Hedberg and Stosick (1952), are given in Table 17. The data from the visual electron diffraction study proved to be correct, although the uncertainties indicated in the early work were large.

It is interesting to compare the molecular geometries of the two aminodiboranes with that of diborane itself. The observed variations

TABLE 17

Geometrical parameters for N,N-dimethylaminodiborane,
$(CH_3)_2NB_2H_5$

	(a)	(b)
r(B$-$B), Å	1.916 ± 0.004	1.92 ± 0.11
r(B$-$H$_t$), Å	$1.191 {+0.010 \atop -0.003}$	
r(B$-$H$_b$), Å	1.365 ± 0.006	
r(B$-$N), Å	1.544 ± 0.010	$1.55_4 \pm 0.02_6$
r(C$-$N), Å	1.488 ± 0.010	$1.48_3 \pm 0.02_9$
$<$H$_t-$B$-$H$_t$	$119.6 \pm 0.5°$	
$<$B$-$H$_b-$B	$89.1 \pm 0.9°$	
$<$B$-$N$-$B	$76.8 \pm 1°$	$76.4 \pm 5.5°$
$<$C$-$N$-$C	$110.0 \pm 1°$	$111.5 \pm 2.5°$
$<$C$-$N$-$B		$116.2 \pm 1.0°$
$<\varepsilon$ (c)	$16.7 \pm 1°$	

(a) Microwave spectroscopy, Cohen and Beaudet (1973).
(b) Visual technique of electron diffraction, Hedberg and Stosick (1952).
(c) The angle between the BH_2 plane and the plane perpendicular to the C_{2v} symmetry axis.

can be accounted for satisfactorily by a decreasing electron deficiency in $(CH_3)_2NB_2H_5$, for example, compared with B_2H_6. This is shown, for instance, by the increase in the relative strength of the B$-$N bond with respect to the B$-$H$_b$ bond in diborane (Hedberg and Stosick, 1952).

Both tetraborane, B_4H_{10}, and pentaborane, B_5H_9, have been studied by the visual technique of electron diffraction (Jones, Hedberg and Schomaker, 1953, and Hedberg, Jones and Schomaker, 1951, respectively). The molecular configurations are shown in Fig. 16. The electron diffraction results are consistent with the crystal-phase X-ray diffraction data for both tetraborane (Nordman and Lipscomb, 1953a, 1953b) and pentaborane (Dulmage and Lipscomb, 1951, 1952). The findings for the structure of pentaborane are also supported by micro-wave spectroscopic data (Hrostowski, Myers and Pimentel, 1952,

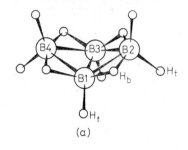

(a)

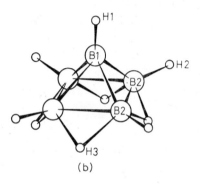

(b)

Fig. 16. (a) Tetraborane; (b) pentaborane

Hrostowski and Myers, 1954). Large uncertainties were indicated for the electron diffraction results. The positions of the hydrogen atoms are particularly ill-determined in the X-ray diffraction studies. Taking all this into consideration, Lipscomb (1963) suggested the following structural parameters as representing the geometries established by the two diffraction techniques:

Tetraborane: $r(B1 - B2) = 1.84$ Å, $r(B1 - B3) = 1.71$ Å, $<B2 - B1 - B4 = 98°$, $r(B - H_t) = 1.19$Å, $r(B1 - H_b) = 1.33$ Å and $r(B2 - H_b) = 1.43$ Å.

Pentaborane: $r(B - B) = 1.80$ Å referring to the base of the tetragonal pyramid and $r(B - B) = 1.69$ Å for the other boron–boron bonds.

The microwave spectroscopic data of Hrostowski and Myers (1954) for pentaborane are consistent with a molecular configuration in which the boron atom skeleton has a tetragonal pyramidal structure

84

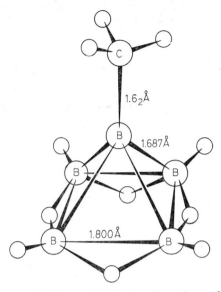

Fig. 17. The molecular configuration of 1-methylpentaborane(9) and the bond distances determined by microwave spectroscopy (Cohen and Beaudet, 1968)

characterized by the following geometrical parameters (the numbering of atoms is shown in Fig. 16):

$$r(B1-B2) = 1.687 \pm 0.005 \text{ Å}$$
$$r(B2-B2) = 1.800 \pm 0.003 \text{ Å} .$$

The molecular model has a C_4 or C_{4v} symmetry. In fact, assuming C_{4v} symmetry and the values $r(B1-H1) = r(B2-H2) = 1.22$ Å, the following parameters could also be obtained:

$$r(B2-H3) = 1.35 \quad \pm 0.02 \text{ Å}$$
$$<B1-B2-H2 = 136°10' \pm 30' .$$

Cohen and Beaudet (1968) applied microwave spectroscopy to 1-*methylpentaborane*(9), 1-$CH_3B_5H_8$, mainly in order to investigate its intramolecular motion. The shape of the molecule and the bond lengths determined are shown in Fig. 17. Almost free rotation around

85

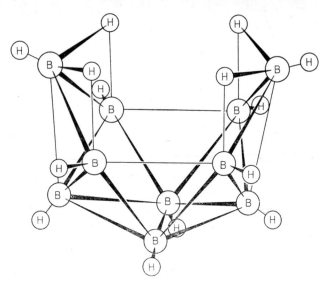

Fig. 18. Decaborane, $B_{10}H_{14}$

the $B-C$ axis was established. We shall discuss some further features of the pentaboranes in connection with the small carboranes (p. 97).

Decaborane, $B_{10}H_{14}$, is the largest boron hydride molecule whose geometry has been studied in the vapour phase. An electron diffraction analysis by Vilkov, Mastryukov and Akishin (1963) was primarily aimed at establishing whether appreciable differences existed between the vapour-phase and crystal-phase molecular structures. The latter was determined by Kasper, Lucht and Harker (1950). It was found, mainly on the basis of comparing various radial distribution curves, that the structures, referring to the two phases, were identical within experimental error. The maxima of the electron diffraction radial distribution appeared at the following positions, with the assignments indicated:

$$1.175 \text{ Å} \quad r(B-H)$$
$$1.78_5 \text{ Å} \quad r(B-B)$$
$$2.81 \text{ Å} \quad r(B \ldots H), r(B \ldots B)$$
$$3.88 \text{ Å} \quad r(B \ldots B), r(B \ldots H).$$

The molecular configuration is presented in Fig. 18.

The carboranes can be derived from the boron hydrides by substituting carbon for one or more of the boron atoms in the boron skeleton. The carbon atom is isoelectronic with the $B-H$ group or the B^- ion. The syntheses of the first representatives of the carboranes were reported in the early 1960s and since then their structure has received much attention. Perhaps their extraordinary stability is the most important chemical behaviour of the carboranes. From the structural aspect, it is also worth mentioning the greater than four coordination of the carbon atoms.

Electron diffraction was applied to elucidate the geometry of 1,5-*dicarba-closo-pentaborane*(5), 1,5-$C_2B_3H_5$, by McNeill, Gallaher, Scholer and Bauer (1973). A trigonal bipyramidal configuration with D_{3h} symmetry was confirmed. The carbon atoms are located at the axial positions. The following bond lengths and bond angles were determined:

$$r(B-C) = 1.556 \pm 0.002 \text{ Å} \quad <B-C-B = 73.05 \pm 0.01°$$
$$r(B-B) = 1.853 \pm 0.002 \text{ Å} \quad <C-B-B = 53.48 \pm 0.01°$$
$$r(C-H) = 1.071 \pm 0.007 \text{ Å} \quad <C-B-C = 93.16 \pm 0.01°$$
$$r(B-H) = 1.183 \pm 0.006 \text{ Å}.$$

The distance from the apical carbon atom to the plane of the three boron atoms is 1.131 Å.

The synthesis of 2-*carbahexaborane*(9), CB_5H_9, was described by Dunks and Hawthorne (1968), who also used NMR studies to establish the presence of two pairs of equivalent boron atoms, three bridge hydrogen atoms and a boron atom at an apical position in the molecule. The corresponding pentagonal pyramidal configuration is shown in Fig. 19. The microwave spectrum of CB_5H_9 was analysed by Cheung and Beaudet (1971). Utilizing a sample with naturally abundant isotopic species, the following boron–boron bond distances were determined:

$$r(B3-B4) = 1.759 \text{ Å}$$
$$r(B4-B5) = 1.830 \text{ Å}$$
$$r(B1-B3) = 1.782 \text{ Å}$$
$$r(B1-B4) = 1.781 \text{ Å}.$$

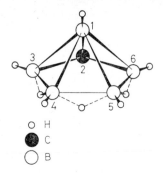

Fig. 19. 2-Carbahexaborane(9), CB_5H_9

The uncertainties are equal to, or slightly larger than 0.007 Å. The microwave spectrum was found to be completely consistent with a configuration suggested by Dunks and Hawthorne (1968). The determination of the position of the carbon atom would have required assumptions regarding the positions of the hydrogen atoms. However, the coordinates obtained for the carbon atom were sensitive to the choice of such assumptions. Isotopic substitution for the carbon atom would be the experimental solution. Assuming the hydrogen atom positions, CNDO/2 calculations have been performed in order to locate the carbon atom. The results of these calculations proved to be insensitive to reasonable variations in the positions of the hydrogen atoms. Accordingly, the following parameters were obtained:

$$r(C-B1) = 1.713 \text{ Å}$$
$$r(C-B3) = 1.530 \text{ Å}$$
$$<H-C-B1 = 130°.$$

The analysis of the NMR spectrum was again useful in establishing the molecular shape and symmetry for 2,3-*dicarba-closo-hexaborane*(6), 2,3-$C_2B_4H_6$, as shown by Shapiro, Keilin, Williams and Good (1963). The detailed molecular geometry was elucidated by the microwave spectroscopic study of Beaudet and Poynter (1970). The molecular configuration and the numbering of atoms are given in Fig. 20a. The following internuclear distances were determined by using thirteen isotopic substitution species:

$r(C2-C3) = 1.540 \pm 0.005$ Å $r(B1-B4) = 1.721 \pm 0.015$ Å
$r(B4-C3) = 1.605 \pm 0.005$ Å $r(B4-B5) = 1.752 \pm 0.005$ Å
$r(B1-C2) = 1.627 \pm 0.015$Å $r(C2-B4) = 2.297 \pm 0.005$Å
$r(B1-B6) = 2.434 \pm 0.005$ Å.

Like 1,5-dicarba-*closo*-pentaborane(5), 1,6-*dicarba-closo-hexabo-rane*(6), $1,6\text{-}C_2B_4H_6$, has no dipole moment and, accordingly, could not be studied by microwave spectroscopy. The molecular geometry of $1,6\text{-}C_2B_4H_6$ was determined by two recent independent electron diffraction studies by Mastryukov, Dorofeeva, Vilkov, Zhigach, Laptev and Petrunin (1973) and McNeill, Gallaher, Scholer and Bauer (1973). The electron diffraction results, which are in agreement with the NMR and IR spectra (Shapiro, Keilin, Williams and Good, 1963), are consistent with a tetragonal bipyramidal model having D_{4h} symmetry, as shown in Fig. 20b. The bond distances reported in the two studies are as follows:

	Mastryukov *et al.* (1973)	McNeill *et al.* (1973)
$r(B-C)$	1.635 ± 0.004 Å	1.633 ± 0.004 Å
$r(B-B)$	1.725 ± 0.012 Å	1.720 ± 0.004 Å
$r(B-H)$	1.15 ± 0.03 Å	1.244 ± 0.012 Å
$r(C-H)$	1.11 Å (assumed)	1.103 ± 0.004 Å

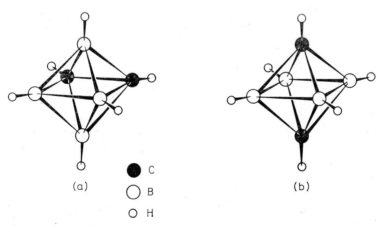

● C
○ B
○ H

Fig. 20. (a) 2,3-Dicarba-*closo*-hexaborane(6), $2,3\text{-}C_2B_4H_6$; (b) 1,6-dicarba-*closo*-hexaborane(6), $1,6\text{-}C_2B_4H_6$

The following bond angles were given by McNeill, Gallaher, Scholer and Bauer (1973):

$$< B2 - C - B3 = 63.55 \pm 0.02°$$
$$< C - B - B \quad = 58.22 \pm 0.02°$$
$$< C - B - C \quad = 83.73 \pm 0.02°$$
$$< B2 - C - B4 = 96.26 \pm 0.02°.$$

As in the case of $1,5\text{-}C_2B_3H_5$, the uncertainties indicated for these bond angles seem to be unrealistically small. The distance from the apical carbon atom to the plane of the four boron atoms is 1.090 Å.

The microwave spectroscopic study of 2,3,4,5-*tetracarbahexaborane*(6), $C_4B_2H_6$, by Pasinski and Beaudet (1973) established the pentagonal pyramidal structure shown in Fig. 21. The following bond distances were obtained:

$$r(C2 - C3) = 1.436 \pm 0.008 \text{ Å}$$
$$r(C3 - C4) = 1.424 \pm 0.007 \text{ Å}$$
$$r(B1 - C2) = 1.709 \pm 0.003 \text{ Å}$$
$$r(B1 - C3) = 1.697 \pm 0.015 \text{ Å}$$
$$r(B6 - C2) = 1.541 \pm 0.007 \text{ Å}$$
$$r(B1 - B6) = 1.886 \pm 0.003 \text{ Å}.$$

For the distances involving the C2 atom, another set of parameters was also obtained due to uncertainty in the choice of the sign of the c coordinate of this atom. The listed values were favoured by the authors. It was noted that because of the relatively short bonds in the base of the pentagonal pyramid, the structure resembles a π-bonded cyclopentadienyl compound with the boron atom in the apical position. The results of this structural determination were consistent with earlier NMR and IR spectroscopic analyses (Onak and Wong, 1970).

The chlorine derivative of the compound previously discussed, *viz.*, 2-*chloro*-1,6-*dicarbahexaborane*(6), $C_2B_4H_5Cl$, possesses a permanent electric dipole moment and its microwave spectroscopic investigation was carried out by McKown and Beaudet (1971). Only a partial structure was obtained using the boron isotopic substitution species. Assuming a molecular model with C_{2v} symmetry, as shown in Fig. 22, the geometrical parameters characterizing the heavy atom skeleton could be determined. In the absence of carbon isotopic substitution species, the values of $r(B - C)$ were calculated by using various assump-

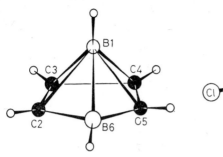
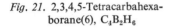

Fig. 21. 2,3,4,5-Tetracarbahexa-
borane(6), $C_4B_2H_6$

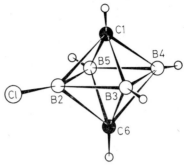

Fig. 22. 2-Chloro-1,6-dicarbahexa-
borane(6), $C_2B_4H_5Cl$

tions. The results were only slightly sensitive to the choice of the assumptions. Thus, the following bond distances and bond angles were found:

$r(B2-Cl) = 1.823 \pm 0.010$ Å $r(C6-B2) = 1.59 \pm 0.04$ Å
$r(B2-B3) = 1.671 \pm 0.010$ Å $r(C6-B4) = 1.61 \pm 0.04$ Å
$r(B3-B4) = 1.702 \pm 0.005$ Å $r(C6-B3) = 1.63 \pm 0.04$ Å

$<B2-B3-B4 = 87.7 \pm 0.5°$
$<B3-B4-B5 = 91.0 \pm 0.2°$
$<B3-B2-B5 = 93.6 \pm 0.5°.$

The NMR and IR spectroscopic analysis of 2,4-*dicarbahepta-borane*(7), $C_2B_5H_7$, provided valuable information concerning the molecular configuration (Onak, Gerhart and Williams, 1963). According to the spectra, no bridge hydrogen atoms occur and there are two pairs of non-equivalent boron atoms in the molecule. A cage-like structure was also proposed but the positions of the carbon atoms were not established. A structure in which there are carbon atoms in the base of the pentagonal pyramid separated from each other by a boron atom was shown to be more favourable by the molecular orbital calculations of Hoffmann and Lipscomb (1962). Other arrangements could not be excluded with confidence, however. One of the aims of the microwave spectroscopic study by Beaudet and Poynter (1965) was to decide between the structures given in Fig. 23. The results definitely supported the correctness of the prediction by Hoffmann

91

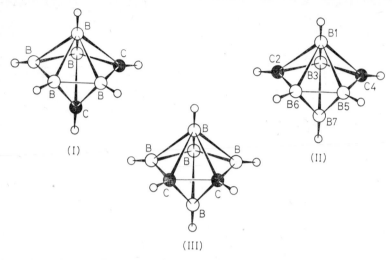

Fig. 23. Three configurations of the $C_2B_5H_7$ molecule. Model (II) was verified by a microwave spectroscopic study (Beaudet and Poynter, 1965)

and Lipscomb (1962). The following bond lengths and bond angles were determined:

$$r(C2-B6) = 1.5627 \text{ Å} \qquad r(B5-B6) = 1.6508 \text{ Å}$$
$$r(C2-B3) = 1.5455 \text{ Å} \qquad r(B1-B5) = 1.8146 \text{ Å}$$
$$r(C2-B1) = 1.7077 \text{ Å} \qquad r(B1-B3) = 1.8177 \text{ Å}$$

$$<C2-B3-C4 = 99°52'$$
$$<B3-C4-B5 = 116°51'$$
$$<C4-B5-B6 = 103°13'$$
$$<B1-B3-B7 = 79°42'.$$

The only uncertainty reported was that for the C2–B3–C4 bond angle ($\pm 30'$).

The following three isomers of *dicarba-closo-dodecaborane*(12), $C_2B_{10}H_{12}$, are known to differ in the positions of the carbon atoms in the icosahedral skeleton (see Fig. 24):

1,2-dicarba-*closo*-dodecaborane(12), or *o*-carborane;
1,7-dicarba-*closo*-dodecaborane(12), or *m*-carborane; and
1,12-dicarba-*closo*-dodecaborane(12), or *p*-carborane.

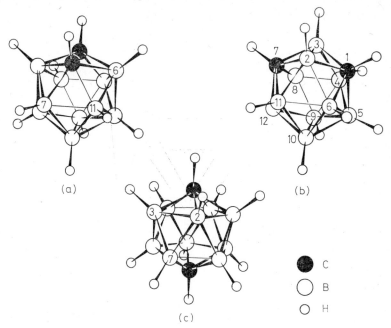

Fig. 24. (a) 1,2-Dicarba-*closo*-dodecaborane(12) or *o*-carborane; (b) 1,7-di-carba-*closo*-dodecaborane(12) or *m*-carborane; (c) 1,12-dicarba-*closo*-dodeca-borane(12) or *p*-carborane (After Bohn and Bohn, 1971)

The vapour-phase molecular geometries of all three isomers have been elucidated by electron diffraction.

The first investigation on *o*-carborane was performed by Vilkov, Mastryukov, Akishin and Zhigach (1965). Because of the complexity of the carborane molecule, a complete determination of its geometry was not even expected. One of the purposes of this study was to establish the relative orientation of the carbon–carbon bond in the heavy atom skeleton. There were two opposing views concerning the molecular configuration of *o*-carborane at the time of the cited work. According to one view (Hoffmann and Lipscomb, 1962; Schroeder, Heying and Reiner, 1963), the C–C bond was closing the boron atom skeleton of decaborane into a regular icosahedron [see (a) in Fig. 24]. Other workers (Zakharkin, Stanko and Brattsev, 1964) suggested a structure in which the carbon–carbon bond is attached to the boron skeleton as if it were an ethylene structure, and the resulting configura-

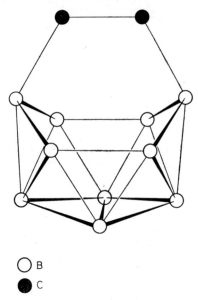

○ B
● C

Fig. 25. The "basket" configuration previously
supposed for *o*-carborane and later discarded

tion resembles a "basket", as shown in Fig. 25. The puzzle of the molecular configuration could not be resolved by the first electron diffraction work on *o*-carborane because equally good agreement was obtained with the experimental data for the following two models: (1) icosahedral configuration with $r(C-C) = 1.68$ Å and $r(B-C) = = 1.70$ Å; and (2) "basket" shape with $r(C-C) = 1.40$ Å and $r(B-C) = 1.60$ Å. The mean length of the boron–boron bonds was found to be 1.76 ± 0.01 Å and it was concluded that the value of $r(B-C)$ lies between 1.60 and 1.70 Å.

Further studies on carborane derivatives verified unambiguously the existence of the icosahedral configuration. The electron diffraction study by Bohn and Bohn (1971) proved this to be true for *o*-carborane itself also. The bond distances determined are given in Table 18. The agreement with the results of the early work is good. The information from the investigation of Bohn and Bohn (1971), however, provides more insight into the structure of *o*-carborane, showing a slight distortion of the icosahedral configuration. They also determined some mean amplitudes of vibration.

TABLE 18

Bond lengths (r_g values) in o-, m- and p-carboranes, from the electron diffraction study of Bohn and Bohn (1971)

	ortho	meta	para
$r(C-C)$, Å	1.653±0.049		
$r(B-C)$, Å	1.711±0.014	1.720±0.009	1.710±0.011
$r(B-B)$, Å*		1.831±0.052	
	1.802±0.013	1.791±0.015	1.792±0.007
	1.789±0.009	1.817±0.013	1.772±0.013
$r(C-H)$, Å	1.15 (assumed)	1.15 (assumed)	1.15 ±0.09
$r(B-H)$, Å	1.216 (assumed)	1.216 (assumed)	1.216±0.021

* The values for the B—B distances are, from top to bottom, in o-carborane for B6—B11 and B7—B11; in m-carborane for B2—B3, B2—B6 and B5—B6 (and B—B bonds more distant from the carbon atoms); in p-carborane for B2—B3 and B2—B7. The numbering of atoms is given in Fig. 24.

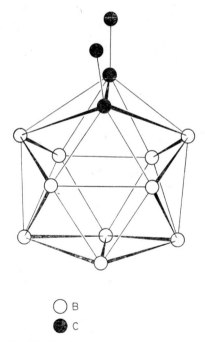

○ B
● C

Fig. 26. The boron–carbon skeleton of
C,C′-dimethyl-o-carborane, $C_2B_{10}H_{10}(CH_3)_2$

The electron diffraction elucidation of the structure of *C,C'-dimethyl-o-carborane*, $C_2B_{10}H_{10}(CH_3)_2$, was reported by Vilkov, Mastryukov, Zhigach and Siryatskaya (1967) and an icosahedral configuration was established. The molecular model is shown in Fig. 26. Only partial information was obtained for the bond distances. The model showing the best agreement with the experimental data included $r(C-C) = 1.70 \pm 0.1$ Å and $r(B-C) = 1.75 \pm 0.05$ Å.

The icosahedral configuration was assumed in the electron diffraction structural analysis of *m-carborane* (or *neocarborane*) by Vilkov, Mastryukov, Zhigach and Siryatskaya (1966). The following bond distances were obtained, mainly on the basis of the experimental radial distribution curves:

$$r(B-B) = r(B-C) = 1.77_5 \pm 0.01 \text{ Å}$$
$$r(B-H) = 1.21 \pm 0.03 \text{ Å}.$$

Hence the boron–boron bond is longer in neocarborane than in analogous boranes and boron–organic compounds. This observation was later verified by Bohn and Bohn (1971), as shown in Table 18.

Beall and Lipscomb (1967) observed a considerable distortion of the icosahedral configuration in an X-ray diffraction study on a crystalline dibromo derivative of neocarborane. The distortion was particularly profound in the vicinity of the carbon atoms. One of the $B-B$ bond lengths was determined to be 1.89 ± 0.04 Å. The electron diffraction data on neocarborane were found to be inconsistent with such a structure (Mastryukov, Vilkov, Zhigach and Siryatskaya, 1969).

The electron diffraction study of *B,B'-diiodoneocarborane*, $C_2B_{10}H_{10}I_2$, was primarily aimed at establishing the positions of the iodine atoms (Vilkov, Khaikin, Zhigach and Siryatskaya, 1968). The assumption of an icosahedral configuration was supported by an examination of several selected models. A mixture of molecular models consisting of 9,10-diiodocarborane-1,7 and 5,12-diiodocarborane-1,7 in equal portions approximated the experimental data well. The presence of 2,9-diiodocarborane-1,7 could not be ruled out, however. The numbering used above is clarified in Fig. 24.

Assuming the icosahedral configuration of the $B_{10}C_2$ skeleton in *C,C'-diiodoneocarborane*, $C_2B_{10}H_{10}I_2$, the $C-I$ bond length was found to be 1.97 ± 0.02 Å by electron diffraction (Mastryukov, Vilkov, Zhigach and Siryatskaya, 1971). The experimental data could be approximated with larger values of $r(C-I)$ only, however, if the refinement of the parameters was carried out for a model with a distorted icosahedral skeleton.

The most accurate determination among the large carborane molecules was achieved for p-carborane, in an electron diffraction study of Bohn and Bohn (1971). This was facilitated by the high symmetry of the molecule. The configuration and the bond lengths determined are presented in Fig. 24 and Table 18, respectively. The distortion of the icosahedral skeleton observed can be visualized as if the regular icosahedron was shrunk by about 10% along the C . . . C axis. Bohn and Bohn (1971) related the distortion to the unusually large H−B2−B7 angle of 130.0 ± 1.8°, which is 8° larger than the value corresponding to the regular structure.

Comparisons of geometrical parameters in analogous molecules are considered below, rather than a discussion of the bonding in carboranes.[*] It is interesting, for instance, to observe the lengthening of the bonds as the coordination number of the skeleton atoms increases. Let us examine first the coordination and bond lengths of the carbon atoms. The $r(B−C)$ values are 1.556 Å and 1.578 Å for the four-coordinated carbon atoms in $1,5\text{-}C_2B_3H_5$ (see p. 87), and in $B(CH_3)_3$ (see Table 4), respectively. The carbon atoms are five-coordinated and $r(B−C) = 1.635$ Å in $1,6\text{-}C_2B_4H_6$. The length of the B−C bond in the chlorine derivative of this compound is not known with sufficient accuracy for this comparison. The carbon atoms are six-coordinated and $r(B−C) = 1.710$ Å in $1,12\text{-}C_2B_{10}H_{12}$. An empirical formula was proposed by Mastryukov, Dorofeeva, Vilkov, Zhigach, Laptev and Petrunin (1973) relating the length of the B−C bond to the n_c coordination number of the carbon atom:

$$r(B−C) = (1.311 + 0.066\,n_c) \text{ Å}.$$

A similar tendency, although less marked, can be observed for the boron–boron bonds, as demonstrated by the data in Table 19. The data discussed above were supplemented by crystal-phase X-ray diffraction results for hexaborane (Cheung and Beaudet, 1971). Only the long B−B bond (1.853 Å) in $1,5\text{-}C_2B_3H_5$ seems to be an exception.

---

[*] The detailed work of Williams (1970) and molecular orbital calculations by Epstein, Koetzle, Stevens and Lipscomb (1970), Koetzle and Lipscomb (1970) and Cheung, Beaudet and Segal (1970) are mentioned for further reference.

TABLE 19

Variations in the lengths of the B—B bonds as the coordination number changes*

Coordination numbers of the boron atoms participating in the bonds	$r(B—B)$, Å		
	5 and 5	5 and 6	6 and 6
Pentaborane		1.69	1.80
1-Methylpentaborane		1.69	1.80
Hexaborane	1.60	1.74	1.75–1.80
2-Carbahexaborane(9)		1.76–1.78	1.78–1.83
1,6-Dicarba-*closo*-hexaborane(6)	1.72₅		
2-Chloro-1,6-dicarbahexaborane(6)	1.70		
2,4-Dicarbaheptaborane(7)	1.65	1.81–1.82	

* For references, see text.

The marked difference from analogous structures observed for this compound was pointed out by McNeill, Gallaher, Scholer and Bauer (1973).

Bohn and Bohn (1971) noted that there is a trend in the variation of the B—B bond lengths in the large carborane molecules. The $r(B—B)$ values referring to interactions nearer to the carbon atoms are larger than those referring to more remote B—B bonds. They explained this phenomenon by the concept of the more electronegative carbon atoms draining electron density away from the neighbouring bonds.

METAL BOROHYDRIDES

Up to the present time, the vapour-phase molecular geometries of beryllium, aluminium and zirconium borohydrides have been studied by electron diffraction. A microwave spectroscopic study of beryllium borohydride has also been attempted. The elucidation of the structures of metal borohydrides is an intricate task owing to the peculiarities of the bonding in these compounds. Again, it has proved fruitful to supplement diffraction techniques with spectroscopic methods that are primarily vibrational analyses. In spite of repeated investigations, some details remain unresolved in some of the determined structures.

Since *beryllium borohydride*, BeB_2H_8, was synthesized for the first time (Burg and Schlesinger, 1940), numerous ideas and models have been advanced for its molecular geometry. Silbiger and Bauer (1946) performed the first electron diffraction study on this compound using the visual technique. On the basis of the positions of the five maxima and the four minima in the electron scattering pattern, a molecular model was proposed with the linear skeleton $H-B-Be-B-H$. The other six hydrogen atoms formed a distorted octahedron around the beryllium atom. This model, with D_{3d} or D_{3h} symmetry, is shown as model (I) in Fig. 27. A model with two hydrogen bridges (D_{2d} symmetry, see (II) in Fig. 27) was favoured, however, on the basis of the infrared spectra and theoretical considerations (for references, see Bauer, 1950; Almenningen, Gundersen and Haaland, 1967, 1968b). Bauer (1950) found that the electron diffraction data were, in fact, consistent with a two-bridge model if the H_b-B-H_b and H_t-B-H_t bond angles and also the $r(B-H_b)$ and $r(Be-H_b)$ bond lengths were chosen to be considerably different. Model (II) in Fig. 27 remained widely accepted until 1967, when Almenningen, Gundersen and Haaland (1967, 1968b) reported the results of a new electron diffraction study using the sector-microphotometer technique. This work showed that the molecular configuration has the three metal atoms at the corners of a roughly equilateral triangle. This configuration had C_{2v} symmetry and is represented by model (III) or (IIIa) in Fig. 27. The study by Almenningen, Gundersen and Haaland (1968b) excluded not only model (I) but also model (IV) of Fig. 27. On the other hand, Cook and Morgan (1969) proposed model (IV) of Fig. 27 on the basis of infrared spectroscopic and mass spectrometric data. They argued that the beryllium atom is strongly bonded in the molecule and that the $BeB_2H_4^+$ ion, which was the most abundant in the 70 eV spectrum, indicates that the parent molecule has a model (IV) configuration. However, as Litzow (1973) pointed out on the basis of the $BeB_2H_4^+$ ion being predominant instead of, $e.g.$, $BeB_2H_3^+$, the existence of model (II) or (III) could be predicted. The "core" of these configurations contains only bridging hydrogen atoms, whereas with model (IV) the appearance of $BeB_2H_3^+$ could be associated.

A more recent vapour-phase infrared spectroscopic study led Cook and Morgan (1970b) to suggest another triangular model with C_{2v} symmetry (see (V) in Fig. 27). The beryllium atom is connected only with bridging hydrogen atoms in this configuration. The presence of some dimeric species was also allowed in order to explain some features of the spectra. Contamination of diborane could have disturbing effects, as was later pointed out by Nibler (1972). The conclusions of another infrared spectroscopic study and the dipole moment measurements favoured the presence of a non-linear $B-Be-B$ skeleton (Nibler and McNabb, 1969). Further support for this arrangement came from the molecular beam deflection experiments of Nibler and Dyke (1970), which established that the molecule is polar. However, attempts to obtain the microwave spectrum of the compound proved to be unsuccessful (Møllendal, 1973). In addition, for some time it was not possible to reproduce the earlier electron diffraction measurements that would have been needed for a more detailed struc-

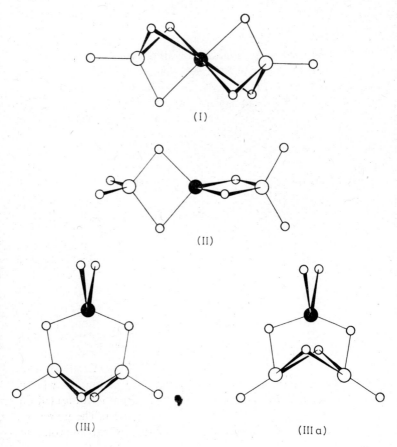

(I)

(II)

(III)   (III a)

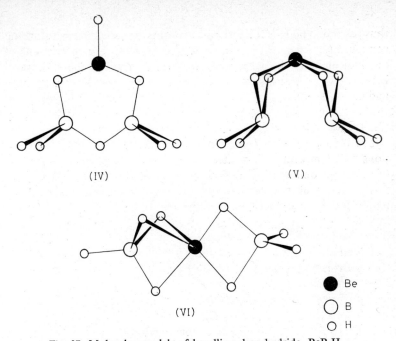

Fig. 27. Molecular models of beryllium borohydride, BeB_2H_8

tural analysis (Haaland, 1973). Meanwhile, the crystal-phase structure was reported by Marynick and Lipscomb (1971) to consist of helical polymers:

thus providing no clue for solving the problem posed by the molecular configuration of the free molecules. Considerable structural changes are indicated by the vibrational spectra, in fact, as the transition between the crystal and vapour phases takes place (Nibler, Shriver and Cook, 1971).

No evidence for the presence of a dimeric species in the vapour was found by the careful infrared and Raman spectroscopic study of Nibler (1972). Certain features of the spectra and the spectral variations with temperature indicated the possible presence of two instead of only one configuration, depending on the conditions of the experiment. Such a phenomenon would also provide an acceptable explanation for the differences between the intensity distributions obtained in various electron scattering experiments. The matrix isolation experiments of Nibler (1972) yielded particularly interesting results. By means of the inert gas matrix, the vapour was cooled to 20 K and the spectrum showed considerable changes that led Nibler to conclude that there was only one configuration present in the matrix while there were two in the vapour. The spectrum of the matrix was consistent with model (I), having C_{3v} symmetry.

Concurrently with the spectroscopic experiments of Nibler (1972) and using samples from the same preparations, a new electron diffraction analysis was performed by Gundersen, Hedberg and Hedberg (1973). Their results did not support the view that the molecule has a triangular heavy atom skeleton. Instead, the configuration originally suggested by Silbiger and Bauer (1946) was confirmed. The electron scattering data were consistent with a model having a linear $B - Be - B$ chain and three bridge hydrogen atoms between the central and terminal atoms. Thus the central atom is six-coordinated. A model with a linear $Be - B - B$ skeleton could not be excluded either, although it seemed to be less probable. It was not possible to distinguish between models (1) having D_{3d} or C_{3v} symmetry. In the latter case, the beryllium atom is not equidistant from the two boron atoms. The following values of the principal geometrical parameters, chosen as averages, and the differences in bond distances and bond angles were given by Gundersen, Hedberg and Hedberg (1973) to characterize the molecular model:

$$r(Be - B) = 1.790 \pm 0.015 \text{ Å}$$
$$0 \leqq \Delta r(Be - B) \leqq 0.10 \text{ Å}$$

$$r(\text{B}-\text{H}_b) = 1.303 \pm 0.012 \text{ Å}$$
$$0 \leqq \varDelta r(\text{B}-\text{H}_b) \leqq 0.12 \text{ Å}$$
$$r(\text{B}-\text{H}_t) = 1.16 \pm 0.04 \text{ Å}$$
$$<\text{H}_b - \text{B} - \text{H}_t = 117.5 \pm 1.2°$$
$$0 \leqq \varDelta <\text{H}_b - \text{B} - \text{H}_t \leqq 10°.$$

Comparing the results of the various experimental studies, it seems probable that different samples were used. In fact, neither of the two main types of molecular configurations, suggested by the vapour-phase investigations, contradict the structure of the polymeric formation observed in the crystal phase (Marynick and Lipscomb, 1971).

The results of *ab initio* calculations by Marynick and Lipscomb (1973) favoured three structures, (II), (VI) and (I) of Fig. 27, and suggested that two or three conformers may co-exist in the gas phase. A similar independent and less complete study by Ahlrichs (1973) arrived at similar conclusions.

The molecular structure of beryllium borohydride is a typical electron deficient structure, in any event, and it shows the rare phenomenon of forming three hydrogen atom bridges between two atoms. Another interesting feature of the model with C_{3v} symmetry is the asymmetric location of the beryllium atom between the two boron atoms. Neither of these two peculiarities is unique, however, in boron stereochemistry. In the bis(cyclopentadienyl)beryllium molecule having C_{5v} symmetry, the beryllium atom is nearer to one of the rings than to the other by 0.43 Å (see pp. 14 and 212). According to Nibler (1972), the electrostatic forces play an important role in forming the actual structure of beryllium borohydride.

The vapour-phase investigations of two beryllium borohydride derivatives should also be mentioned. Cook and Morgan (1970a) reported on the spectroscopic properties of *methylberyllium borohydride*, CH_3BeBH_4. The presence of a mixture of monomeric and dimeric species was established in the vapour phase. The suggested molecular configurations are presented in Fig. 28. The structure of the monomer is consistent with the molecular geometry of dimethylberyllium, $Be(CH_3)_2$, as determined by electron diffraction (Almenningen, Haaland and Morgan, 1969) and vibrational spectroscopy (Kovar and Morgan, 1969). The molecular configuration of the dimer, on the other hand, resembles that of the trimethylaluminium dimer (cf., p. 108).

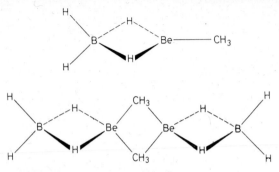

Fig. 28. Suggested molecular configurations for monomeric and dimeric methylberyllium borohydride (Cook and Morgan, 1970a)

The geometrical parameters of *cyclopentadienylberyllium borohydride*, $C_5H_5BeBH_4$, have been refined for two models in an electron diffraction structural analysis by Drew, Gundersen and Haaland (1972). There are two bridge hydrogen atoms in one of the models and the $BeBH_4$ fragment possesses C_{2v} symmetry:

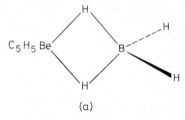

(a)

The beryllium and boron atoms are connected by three bridge hydrogen atoms in the other model and the $BeBH_4$ fragment has C_{3v} symmetry:

$$C_5H_5Be \overset{\displaystyle H}{\underset{\displaystyle H}{\overline{H}}} B \text{———} H$$

(b)

Both models are consistent with the electron diffraction data. The parameters obtained for the C_5H_5Be part (see Table 52) proved to be insensitive to the choice of the overall molecular model. The following bond lengths (Å) were obtained for the $BeBH_4$ group in the case of the two refinement schemes:

	(a)	(b)
$r(Be-B)$	1.88 ± 0.01	1.89 ± 0.01
$r(B-H_t)$	1.17 ± 0.03	1.16 ± 0.02
$r(B-H_b)$	1.29 ± 0.05	1.28 ± 0.03
$r(Be-H_b)$	1.78 ± 0.09	1.70 ± 0.05

Thus the differences between the corresponding data referring to the two models do not exceed the error limits. It is interesting that the $Be-B$ bond is considerably longer than that in beryllium borohydride (Gundersen, Hedberg and Hedberg, 1973).

The first electron diffraction study on *aluminium borohydride*, $Al(BH_4)_3$, was performed by Beach and Bauer (1940) using the visual technique. They demonstrated that the aluminium atom is bonded

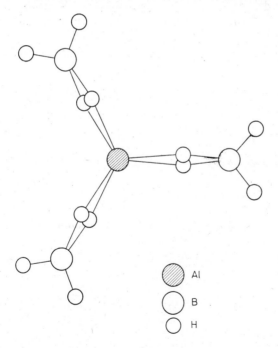

Fig. 29. The molecular configuration of aluminium borohydride, $Al(BH_4)_3$

to all three boron atoms and all of the B–Al–B angles are close to 120°. The position of the hydrogen atoms was established by an infrared spectroscopic investigation by Price (1949), who concluded that a bridge structure similar to that of diborane is formed. The molecular configuration is shown in Fig. 29. Using the up-to-date technique of electron diffraction, Almenningen, Gundersen and Haaland (1968a) examined the molecular symmetry and intramolecular motion, in addition to making an accurate determination of the geometrical parameters. Two models had to be considered in detail. In one of them, the line joining the bridge hydrogen atoms is perpendicular to the AlB_3 plane. The six bridge hydrogen atoms form a trigonal prism around the aluminium atom and, accordingly, the molecular symmetry is D_{3h}. By turning the BH_4 tetrahedra around the Al–B axes, a molecular model with D_3 symmetry can be obtained. The electron diffraction data showed good general agreement with the D_{3h} model except for some deviations in the region of the large interatomic distances on the radial distribution curves. These deviations could be reduced by a slight distortion of the D_{3h} symmetry. The torsional angle of the BH_4 tetrahedra around the Al–B axes was found to be $17.2 \pm 2.4°$ in the least-squares refinement. However, the authors did not rule out the model with D_{3h} symmetry, as the appearance of the D_3 model may only be a consequence of shrinkage effects. The following bond distances and bond angles were determined:

$r(Al–B)$ $= 2.143 \pm 0.003$ Å $<H_b–Al–H_b = 73.4 \pm 0.8°$
$r(Al–H_b)$ $= 1.801 \pm 0.006$ Å $<H_b–B–H_b = 114.0 \pm 0.2°$
$r(B–H_b)$ $= 1.283 \pm 0.012$ Å $<H_t–B–H_t = 116.2 \pm 2.2°$
$r(B–H_t) = 1.196 \pm 0.012$ Å.

The electron diffraction elucidation of the molecular structure of *zirconium borohydride*, $Zr(BH_4)_4$, was performed by two groups independently at about the same time. Except for some fine details, there is good general agreement between the findings of the two vapour-phase studies by Plato and Hedberg (1970) and Spiridonov and Mamaeva (1969). The vapour-phase data are also consistent with the results of a low-temperature crystal-phase X-ray diffraction investigation by Davies, Wallbridge, Smith and James (1973). The results of the more complete electron diffraction study are summarized in Table 20. The principal difference in the interpretation of the electron

TABLE 20

Structural parameters for $Zr(BH_4)_4$ determined by the electron diffraction investigation of Plato and Hedberg (1970)

	r_a	l_a
Zr−B, Å	2.308±0.010	0.045±0.010
Zr−H$_t$, Å	2.211±0.04	0.139±0.055
B−H$_t$, Å	1.18 ±0.12	0.07 ±0.12
B−H$_b$, Å	1.272±0.05	0.072±0.044
Zr...H$_t$, Å	3.48 ±0.12	0.18 ±0.20
B...B, Å	3.770±0.016	0.10 ±0.04
<H$_b$−B−H$_b$	108.4±2.1°	
Torsional angle of BH$_4$ groups	38°	

diffraction data concerns the type of bonding rather than the molecular geometry. There are three bridge hydrogen atoms connecting the zirconium atom with each boron atom according to Plato and Hedberg (1970), as illustrated by model (a) in Fig. 30. On the other hand, Spiridonov and Mamaeva (1969) concluded that the central atom and the BH_4 groups may be linked by zirconium–boron bonds [see model (b) in Fig. 30]. The available physical evidence, however, does not indicate unambiguously whether a chemical bond is formed between the two atoms or not. The vapour-phase infrared spectra (Bird and Churcill, 1967) are completely consistent with the electron diffraction results in that the molecule of $Zr(BH_4)_4$ has high sym-

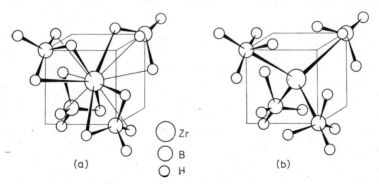

Fig. 30. The molecular configuration of zirconium borohydride, $Zr(BH_4)_4$, (a) after Plato and Hedberg (1970); (b) after Spiridonov and Mamaeva (1969)

metry and the molecular geometry is characterized by three bridge hydrogen atoms forming the linkage between the central atom and each of the boron atoms.

METAL ALKYLS

It is a common phenomenon that metal alkyl molecules* are linked together by forces similar to those which hold together the two units of the diborane molecule. In the light of the interesting bonding properties of the metal alkyls, it is surprising that their vapour-phase molecular geometries have been studied only occasionally.

The dimeric methylberyllium borohydride (Fig. 28), which has already been discussed, also belongs, in fact, to this class of compounds.

The molecular structure of *trimethylaluminium dimer*, $Al_2(CH_3)_6$, has undergone detailed studies. The electron diffraction data are consistent with a D_{2h} symmetry model:

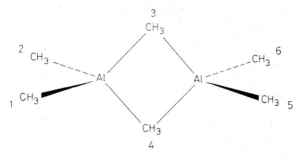

as was shown by Almenningen, Halvorsen and Haaland (1971). The bond lengths and bond angles determined are given in Table 21. The vapour-phase results are in good agreement with the findings of a crystal-phase X-ray diffraction study by Vranka and Amma (1967). It is interesting that negative (-0.115 ± 0.080 Å) and positive (0.122 ± 0.026 Å) shrinkages were observed in the vapour phase for the distances C1 . . . C5 and C1 . . . C6, respectively. This was interpreted as a result of the torsional motion of the terminal AlC_2 groups around the aluminium—aluminium axis. Almenningen, Halvorsen and

* Typical representatives are the aluminium alkyl compounds, for example (Mole and Jeffry, 1972).

TABLE 21

Bond lengths and bond angles in trimethylaluminium dimer,
$Al_2(CH_3)_6$, molecules

	(a)	(b)
$r(C-H)$, mean, Å	1.117 ± 0.002	
$r(Al-C1)$, Å	1.957 ± 0.003	1.956 ± 0.002, 1.949 ± 0.002
$r(Al-C3)$, Å	2.140 ± 0.004	2.125 ± 0.002, 2.123 ± 0.002
$r(Al-Al)$, Å	2.619 ± 0.005	2.606 ± 0.002
$<Al-C-Al$	$75.5 \pm 0.1°$	$75.7 \pm 0.1°$
$<C1-Al-C2$	$117.3 \pm 1.5°$	$123.2 \pm 0.1°$

(a) Vapour-phase structure by electron diffraction (r_a parameters), Almenningen, Halvorsen and Haaland (1971).

(b) Crystal-phase X-ray diffraction results, Huffman and Streib (1971).

Haaland (1971) ascribed the differences between the respective structural parameters from the electron diffraction and X-ray diffraction
studies to the influence of lattice forces in the crystal and the absence
thereof in the vapour. The electron-deficient bridge structure of
trimethylaluminium dimer can be considered to be completely analogous to that of diborane. The only difference is that in $Al_2(CH_3)_6$
the methyl carbon atoms are the bridges. It is the latter view, however,
which is still debated. According to Byram, Fawcett, Nyberg and
O'Brien (1970), the monomer units are held together by two-electron
bonds:

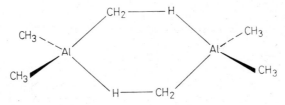

This conclusion was based on a refinement of the data from a paper
by Vranka and Amma (1967). As was pointed out by Cotton (1970),
however, these data were not accurate enough to justify the further
refinement performed by Byram, Fawcett, Nyberg and O'Brien
(1970). The recent X-ray diffraction study of Huffman and Streib
(1971), on the other hand, yielded accurate data concerning the position of the hydrogen atoms and excluded the possibility of the presence of bridging hydrogen atoms. Some of the geometrical param-

eters determined by Huffman and Streib (1971) for the crystalline trimethylaluminium dimer are presented in Table 21. The gaseous electron diffraction evidence for the three-centre bond in $Al_2(CH_3)_6$ is not decisive. Although the sample must be in the crystal phase, a neutron diffraction investigation would be most desirable as it is the most suitable technique for determining the position of the bridge hydrogen atoms.

The molecular geometry of *dimethylaluminium hydride dimer*, $[HAl(CH_3)_2]_2$, was elucidated by the electron diffraction study of Almenningen, Anderson, Forgaard and Haaland (1971, 1972). For a model with D_{2h} symmetry:

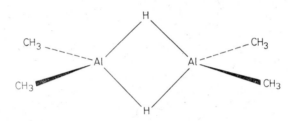

the following geometrical parameters were obtained:

$r(C-H) = 1.117 \pm 0.005$ Å $r(Al-C) = 1.947 \pm 0.003$ Å
$r(Al-H) = 1.676 \pm 0.019$ Å $r(Al-Al) = 2.617 \pm 0.006$ Å

$<C-Al-C = 118.5 \pm 0.9°$
$<Al-H-Al = 102.5 \pm 1.6°$.

It is of interest to compare the structural parameters of $Al_2(CH_3)_6$ and $[HAl(CH_3)_2]_2$. The values for both $r(Al-Al)$ and $l(Al-Al) = 0.079$ Å are the same in the two molecules. The $Al-Al$ distance is relatively small,* indicating a strong direct linkage, apparently a general characteristic of electron-deficient molecules (Levison and Perkins, 1970). The other bonds are relatively long and, accordingly, weaker. The intra-ring bond angles in the two molecules are considerably different, as a consequence of the difference in the bond distances $r(Al-C)$ and $r(Al-H)$.

---

* Twice the covalent radius would be 2.52 Å (Pauling, 1960).

HALOGEN BRIDGING COMPLEXES

We begin this section by mentioning three important reference papers. Bauer and Porter (1964) surveyed the literature concerning structural and thermochemical studies on vapour-phase metal halides up to 1963. Novikov and Gavryuchenkov (1967) reviewed the behaviour of high-temperature vapours of mixed halides. Szpiridonov (1972) summarized the high-temperature electron diffraction investigations performed at the Moscow State University on vapour-phase complex halides.

Studies of the vapour-phase complex halides are neither less interesting nor less informative than the investigations on condensed-phase species. This is well demonstrated by the rapid increase in the associated literature. The structural information on vapour-phase complex halides is of great importance both for theory (chemical bonding) and practice (see, *e.g.*, thermodynamic calculations needed for metallurgy). Polymerization processes produce various complex molecules in the vapours of halides. The formation of the halogen bridging bond is usually explained by the electron-donating properties of the halogen atoms towards the metal atoms:

According to the experimental data, the bridging bonds in the ring are equivalent. Structural studies on the vapour-phase halides often require experimental work at elevated temperatures.

The discussions below on dimeric, trimeric and tetrameric halides and also on mixed halide complexes also include structural information on the monomeric species.

111

Among the halides of the alkali metals, the tendency for association to occur increases from the caesium halides towards the lithium halides (see, *e.g.*, Klemperer and Rice, 1957; Berry and Klemperer, 1957; Milne, Klein and Cubicciotti, 1958; Berkowitz and Chupka, 1958; Eisenstadt, Rothberg and Kusch, 1958; Bauer, Diner and Porter, 1958; Pugh and Barrow, 1958). So far, the molecular geometries have been elucidated only for the lithium salts (Bauer, Ino and Porter, 1960; Akishin and Rambidi, 1960; Akischin and Rambidi, 1960) and thallium fluoride (Solomonik, Zasorin, Girichev and Krasnov, 1974).

The results of electron diffraction studies on lithium halides are consistent with a planar ring model:

The geometric parameters are presented in Table 22. The planarity of the ring has been assumed rather than determined in these studies. The uncertainty is caused by the relatively weak electron-scattering properties of the lithium atom and the large amplitude of vibration of the Li . . . Li atomic pair, which hinders an accurate determination of the Li . . . Li distance. Infrared spectroscopic studies, including a matrix isolation investigation (Snelson, 1967), support the assumption of the planarity of the ring of Li_2X_2. Further evidence was provided by a molecular beam deflection experiment (Büchler, Stauffer and Klemperer, 1964b; cf., Table 1), establishing the absence of a permanent electric dipole moment.

The existence of dimeric species in the vapours of other alkali metal halides was also indicated by electron diffraction evidence. The bond distances determined by the early visual electron diffraction studies for what were supposed to be monomeric species are greater than those obtained by microwave spectroscopy, as shown in Table 23. The differences cannot be accounted for simply by the difference in the physical significance of the parameters produced by the two

112

TABLE 22

Geometrical parameters for vapour-phase lithium halides

		$r(Li-X)$, Å	$<X-Li-X$
LiCl	(a)	2.02067 ± 0.00006	
Li_2Cl_2	(b)	2.23 ± 0.03	$108 \pm 4°$
LiBr	(c)	2.17042 ± 0.00004	
Li_2Br_2	(d)	2.35 ± 0.02	$110 \pm 4°$
LiI	(c)	2.39191 ± 0.00004	
Li_2I_2	(d)	2.54 ± 0.02	$116 \pm 4°$

(a) Microwave spectroscopy, r_e parameter, Lide, Cahill and Gold (1964).

(b) Visual technique of electron diffraction, Bauer, Ino and Porter (1960).

(c) Microwave spectroscopy, r_e parameter; for references see Lide, Cahill and Gold (1964).

(d) Visual technique of electron diffraction, Akishin and Rambidi (1960) and also Akischin and Rambidi (1960).

TABLE 23

Bond distances (Å) in sodium, potassium, rubidium and caesium halides determined for the vapour phase by electron diffraction (ED) and microwave spectroscopy (MW)

	$r(ED)*$	$r(MW)**$	$r(ED)-r(MW)$
NaF	2.02	1.92593 ± 0.00006	0.09
NaCl	2.45	2.3606 ± 0.0001	0.09
NaBr	2.57	2.50201 ± 0.00004	0.07
NaI	2.80	2.71143 ± 0.00004	0.09
KF	2.20	2.17144 ± 0.00005	0.03
KCl	2.70	2.6666 ± 0.0001	0.03
KBr	2.86	2.82075 ± 0.00005	0.04
KI	3.09	3.04781 ± 0.00005	0.04
RbF	2.32	2.26554 ± 0.00005	0.05
RbCl	2.83	2.78670 ± 0.00006	0.04
RbBr	2.97	2.94471 ± 0.00005	0.03
RbI	3.17	3.17684 ± 0.00005	-0.01
CsF	2.33	2.3453 ± 0.0001	-0.02
CsCl	2.91	2.9062 ± 0.0001	0.00
CsBr	3.07	3.07221 ± 0.00005	0.00
CsI	3.31	3.31515 ± 0.00006	-0.01

* Visual technique, Akishin, Rambidi, Kuznetsov and Matrosov (1957); Akishin and Rambidi (1958a); Akishin and Rambidi (1958b); Akishin and Rambidi (1959).

** r_e parameters; for references see, e.g., Lide, Cahill and Gold (1964).

techniques. The rotational transitions of the monomers can be assigned unambiguously in the rotational spectra even in the presence of both monomers and dimers in the vapours as a consequence of the simplicity and high resolution of the spectra. On the other hand, superposition of the electron scattering from the monomer and dimer is obtained in the electron diffraction experiment. Accordingly, an effective bond distance is yielded by the structural analysis, corresponding to the relative abundance of monomers and dimers if the composition of the vapour is not taken into consideration. Thus, *e.g.*, the vapour of the sodium halides contains approximately 35% dimers (Miller and Kusch, 1956), and the bond distances determined by electron diffraction are considerably different from those obtained by spectroscopy. On the other hand, dimerization of the caesium compounds could not be detected (Miller and Kusch, 1956) and the bond distances determined by the two techniques are almost identical. The correlation between the monomer : dimer ratio and the effective bond distance, $r_{eff.}$, is illustrated by the curve shown in Fig. 31. The values of $r_{eff.}$ can be calculated from the theoretical distribution curves corresponding to different vapour compositions. On the basis of correlation curves such as that shown in Fig. 31, the following conclusions were drawn by Akishin and Rambidi (1960): (1) the difference between the values of $r_{eff.}$ and $r_{dim.}$ (the bond distance in the dimer) is small even in the presence of a considerable amount of monomers; (2) if the value of $r_{mon.}$ (the bond distance in the monomer) and the composition of the vapour are known, the value of $r_{dim.}$ can be obtained. The bond distances in the monomers have been measured by microwave spectroscopy with great accuracy (see Tables 22 and 23). Accordingly, the value of $r_{dim.}$, or, in the knowledge of both $r_{mon.}$ and $r_{dim.}$, the composition of the vapour, i.e., the degree of dimerization, could be evaluated (Akishin, Rambidi and Spiridonov, 1967). With up-to-date techniques of structural analysis, it is expected that both the bond distances and the degree of dimerization can be determined.

It is interesting to note that the metal–halogen bond in the dimer is longer than that in the monomer in the alkali metal halides. A further lengthening of the bond was observed for lithium compounds in the crystalline phase (see Bauer, Ino and Porter, 1960): lithium chloride 2.57 Å; lithium bromide 2.75 Å; and lithium iodide 3.00 Å.

Dimeric species prevail in the vapours of thallium(I) halides (Cubicciotti, 1970; Keneshea and Cubicciotti, 1965, 1967). These com-

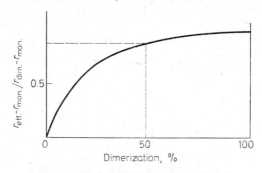

Fig. 31. The variation of the quantity
$(r_{\text{eff.}} - r_{\text{mon.}})/(r_{\text{dim.}} - r_{\text{mon.}})$ as a function of the degree
of dimerization (Akishin and Rambidi, 1960)

pounds have been the subject of numerous spectroscopic studies (see, *e.g.*, Brom and Franzen, 1971, and references therein). Of the various possibilities, the following three models have been studied in more detail:

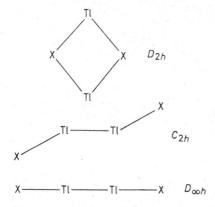

The electron diffraction data on thallium(I) fluoride (Solomonik, Zasorin, Girichev and Krasnov, 1974) showed an 80% relative abundance of dimers in the vapour. The results are consistent with a planar irng configuration (D_{2h} symmetry). The following parameters were given as final results:

$r(\text{Tl}-\text{F})\ \ = 2.29 \pm 0.02\ \text{Å}$ $l(\text{Tl}-\text{F})\ \ = 0.09_5 \pm 0.01_5\ \text{Å}$
$r(\text{Tl}\ldots\text{Tl}) = 3.68 \pm 0.01\ \text{Å}$ $l(\text{Tl}\ldots\text{Tl}) = 0.14 \pm 0.01\ \text{Å}.$

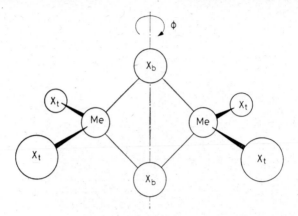

Fig. 32. The configuration of Me_2X_6 molecules (Me = Al, Ga, In or Fe; X = Cl, Br or I). Large amplitude torsional motion has been observed in some molecules around the axis indicated

On the other hand, according to a matrix isolation infrared spectroscopic investigation (Brom and Franzen, 1971), the Tl_2X_2 molecules have a linear configuration.

There is experimental evidence for the existence of dimers of alkaline earth metal halides (Berkowitz and Marquart, 1962), but no structural information has yet been obtained.

Dimeric Me_2X_6 molecules with bridge structures are formed in the vaporization of trichlorides, tribromides and triiodides of aluminium, gallium and indium. The molecules consist of two tetrahedra sharing a common edge (D_{2h} symmetry), as shown in Fig. 32. The Me atoms (Al, Ga or In) are located in the centre of the tetrahedron and the X atoms (Cl, Br or I) are found at the apexes of the tetrahedra. The aluminium and gallium compounds have often been studied (Akishin, Rambidi and Zasorin, 1959; Palmer and Elliot, 1938; Stevenson and Schomaker, 1942; Akishin, Naumov and Tatevskii, 1959; Zasorin, 1965; Butaev, 1970; Shen, 1973). Some of the structural data are summarized in Tables 24 and 25. The presence of large amplitude torsional motion around the imaginary axis connecting the bridging halogen atoms (see Fig. 32) was pointed out by Shen (1973).

In the vapours of aluminium iodide and gallium bromide, both monomeric and dimeric molecules are present (Shen, 1973). For the

TABLE 24

Vapour-phase molecular geometries of aluminium chloride, aluminium bromide and aluminium iodide dimers

	Al_2Cl_6		Al_2Br_6		Al_2I_6
	(a)	(b)	(a)	(c)	(a)
$r(Al-X)_t$, Å	2.065 ± 0.003	2.07 ± 0.01	2.222 ± 0.005	2.22 ± 0.02	2.449 ± 0.093
$r(Al-X)_b$, Å	2.252 ± 0.003	2.26 ± 0.02	2.414 ± 0.007	2.38 ± 0.02	2.634 ± 0.030
$<X_t-Al-X_t$	$123.4 \pm 1.6°$	$120 \pm 1°$	$122.8 \pm 3.3°$	$118 \pm 3°$	$115.0 \pm 7.4°$
$<X_b-Al-X_b$	$91.0 \pm 1.6°$	$91 \pm 1°$	$92.3 \pm 0.9°$	$82 \pm 3°$	$99.6 \pm 4.5°$
ϕ	$23.4 \pm 6.0°$		$23.3 \pm 14.0°$		$25.0°$ (assumed)

(a) Electron diffraction, Shen (1973). The parameter ϕ is the root-mean-square amplitude of bending motion around an imaginary axis drawn between the two bridging halogen atoms.

(b) Electron diffraction, Butaev (1970).

(c) Electron diffraction, Akishin, Rambidi and Zasorin (1959a).

TABLE 25

Vapour-phase molecular geometries of gallium chloride and gallium bromide dimers

| | Ga_2Cl_6 | | Ga_2Br_6 | |
	(a)	(b)	(a)	(b)
$r(Ga-X)_t$, Å	2.099±0.002	2.09±0.02	2.245±0.003	2.25±0.02
$r(Ga-X)_b$, Å	2.300±0.003	2.29±0.02	2.446±0.009	2.35±0.02
$<X_t-Ga-X^t$	124.6±1.8°	112±3°	128.1±3.0°	110±3°
$<X_b-Ga-X_b$	88.3±0.8°	91±3°	91.1±2.2°	93±3°
ϕ	20.5±3.3°		25.0° (assumed)	

(a) See Table 24 (a).
(b) Electron diffraction, Akishin, Naumov and Tatevskii (1959).

monomers, D_{3h} symmetry was established with the following geometrical parameters for $GaBr_3$: $r(Ga-Br) = 2.243 \pm 0.040$ Å and $<Br-Ga-Br = 120.0 \pm 1.2°$. The structural data for monomeric aluminium chloride and iodide are presented in Table 7.

A bridge structure with D_{2h} symmetry was established for the dimers of indium halides by the visual technique of electron diffraction (Stevenson and Schomaker, 1942). The following average values were obtained for the bond distances:

$$In_2Cl_6 \qquad r(In-Cl) = 2.46 \text{ Å}$$
$$In_2Br_6 \qquad r(In-Br) = 2.58 \text{ Å}$$
$$In_2I_6 \qquad r(In-I) = 2.76 \text{ Å}.$$

The vapour-phase molecular geometry of *dimethylaluminium chloride* dimer, $[(CH_3)_2AlCl]_2$, was elucidated in an up-to-date electron diffraction investigation by Brendhaugen, Haaland and Novak (1974). The configuration possessing D_{2h} symmetry

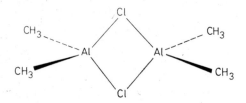

is characterized by the following bond distances and bond angles:

$$r(Al-C) = 1.935 \pm 0.004 \text{ Å}$$
$$r(Al-Cl) = 2.303 \pm 0.003 \text{ Å}$$
$$<C-Al-C = 126.9 \pm 0.8°$$
$$<Cl-Al-Cl = 89.4 \pm 0.5°.$$

Both this compound and *dimethylaluminium bromide* dimer, $[(CH_3)_2AlBr]_2$, had been studied earlier by the visual technique (Brockway and Davidson, 1941). The molecular parameters for the bromine derivative from this early study are: $r(Al-C) = 1.98 \pm 0.08 \text{ Å}$, $r(Al-Br) = 2.42 \pm 0.03 \text{ Å}$, $<C-Al-C = 122.5 \pm 7.5°$ and $<Br-Al-Br = 90 \pm 3°$.

The terminal bonds appear to be shorter than the bridging bonds in all Me_2X_6 molecules. A similar difference was observed for the $Al_2(CH_3)_6$ molecule (cf., Table 21). Hence, the phenomenon is not a consequence of the difference in the bonding peculiarities between the two types of compounds (Wade, 1968; 1972). Remarkable constancy is demonstrated by the differences and ratios of terminal and bridging bond distances:

	Cl	Br	CH_3
$r(Al-X)_b - r(Al-X)_t$	0.19	0.16	0.18
$r(Al-X)_t/r(Al-X)_b$	0.92	0.93	0.91

Wade (1972) examined in detail the correlation between the difference of the terminal and bridging bond distances and the bond energy. It was also concluded that the bridging bonds are always subjected to stronger repulsive interactions than are the terminal bonds. Hence, even in the case of formally equal bond order, the bridging bond is weaker than the terminal bond. This difference then increases as the bond order of the terminal bond becomes higher than that of the bridging bond. The formal bond order of the terminal bond in the electron-deficient bridging structures is 1.0, while that of the bridging bond is 0.5. For the halides, where there is a dative bond from the halogen atom towards the metal atom, the bond order of the terminal bond is greater than unity and the bridging bond can be considered to be a single bond.

The terminal Al–Cl bond in dimeric aluminium chloride has almost the same length as the bond in the monomer (see Table 7). On the other hand, the terminal Al–C bond in dimeric dimethyl-

119

aluminium chloride is considerably shorter than the analogous bond in monomeric or dimeric trimethylaluminium (see Tables 7 and 21). Another important observation is that the $C-Al-C$ bond angle in $[(CH_3)_2AlCl]_2$ is larger than the C_t-Al-C_t bond angle in dimeric trimethylaluminium. Both phenomena are attributed to greater s-orbital participation for $[(CH_3)_2AlCl]_2$ than for $Al_2(CH_3)_6$ in forming the $Al-C_t$ bond (Brendhaugen, Haaland and Novak, 1974). The vapour-phase data can be supplemented by the results of an X-ray diffraction study (Allegra, Perego and Immirzi, 1963) on crystalline methylaluminium dichloride:

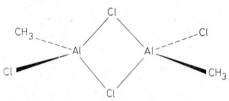

where $r(Al-Cl_t) = 2.05 \pm 0.01$ Å, $r(Al-Cl_b) = 2.26 \pm 0.01$ Å, $r(Al-C)=1.93\pm0.03$ Å, $<Al-Cl_b-Al=91.1\pm0.5°$, $<Cl_t-Al-Cl_b= = 105.8$ and $107.6 \pm 0.5°$, $<C-Al-Cl_b = 110.4$ and $112.7 \pm 1.5°$ and $<C-Al-Cl_t = 124.5 \pm 1.5°$. The terminal substituents are again in *anti*-positions and this observation is supported by vibrational spectroscopic evidence (Mach, 1965; Weidlein, 1969). The overall structure is consistent with the vapour-phase geometries of Al_2Cl_6 and $[(CH_3)_2AlCl]_2$. The bridging $Al-Cl$ bonds in both Al_2Cl_6 and $(CH_3AlCl_2)_2$ are shorter than in $[(CH_3)_2AlCl]_2$.

The vapour-phase and crystal-phase (Renes and MacGillavry, 1945) data on Al_2Br_6 are in agreement concerning the shape of this molecule. A quantitative comparison between the two sets of data is not relevant, however, because of the large uncertainties. The structure of aluminium chloride in the crystal phase is different. There are no Al_2Cl_6 molecules and each aluminium atom is surrounded by six chlorine atoms (Bigelow, 1969).

The molecular geometry of *iron(III) chloride* dimer, Fe_2Cl_6, is analogous to that of the aluminium compound according to the electron diffraction study of Zasorin, Rambidi and Akishin (1963). The following parameters were determined:

$$r(\text{Fe} - \text{Cl})_t = 2.11 \pm 0.03 \text{ Å}$$
$$r(\text{Fe} - \text{Cl})_b = 2.28 \pm 0.03 \text{ Å}$$
$$< \text{Cl}_t - \text{Fe} - \text{Cl}_t = 128 \pm 3°$$
$$< \text{Cl}_b - \text{Fe} - \text{Cl}_b = 92 \pm 3°.$$

The value of the length of the terminal bond agrees, within experimental error, with that of the bond distance in monomeric $FeCl_3$ (2.14 ± 0.01 Å) determined by an electron diffraction study on over-heated iron(III) chloride vapour (Rambidi and Zasorin, 1964).

TRIMERS

As studies on the vapours of halides increase in number, more and more trimeric halides with bridge structures are becoming known. For example, there are Li_3F_3 molecules also present in the vapour of lithium fluoride in addition to LiF and Li_2F_2 molecules, according to a matrix isolation infrared spectroscopic study (Snelson, 1967). The molecular geometry, however, has been determined so far for two trimers only, namely copper(I) chloride and rhenium(III) bromide.

Wong and Schomaker (1957) established D_{3h} symmetry for copper(I) chloride trimer, Cu_3Cl_3, molecules. Their visual electron diffraction study yielded also the value of 2.160 ± 0.015 Å for the bridging Cu − Cl bond distance, with bond angles of Cu − Cl − Cu = 90° and Cl − Cu − Cl = 150°. Copper(I) bromide and iodide had been studied earlier (see Sutton, 1958) without producing sufficient information on the composition of the vapour. Accordingly, only the bond distances $r(\text{Cu} - \text{Br}) = 2.25$ Å and $r(\text{Cu} - \text{I}) = 2.40$ Å could be determined then. Later, the data for these compounds were interpreted in terms of a structure analogous to that of Cu_3Cl_3 (Bauer and Porter, 1964).

The sector-microphotometer technique of electron diffraction was employed in a study of *rhenium(III) bromide* trimer, Re_3Br_9 by Ugarov, Vinogradov, Zasorin and Rambidi (1971). The molecular

121

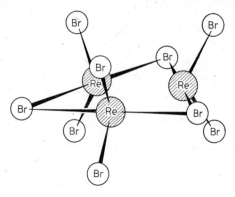

Fig. 33. The molecular configuration of Re_3Br_9

configuration with D_{3h} symmetry is shown in Fig. 33. The bond distances and bond angles were found to be:

$$r(Re-Re) = 2.46 \pm 0.02 \text{ Å}$$

bridging $\quad r(Re-Br) = 2.55 \pm 0.02 \text{ Å}$

terminal $\quad r(Re-Br) = 2.43 \pm 0.02 \text{ Å}$

$$< Re-Br-Re = 57 \pm 2°$$

$$< Br_t-Re-Br_t = 142 \pm 6°.$$

The rhenium–rhenium internuclear distance is strikingly short, much shorter than twice the covalent radius (1.283 Å, Pauling, 1960). Thus, in addition to the halogen bridges, chemical bonding between the rhenium atoms seems to exist.

There are almost exclusively trimeric Re_3X_9 species in the vapours of rhenium bromide and chloride in the temperature range 240–350°C according to mass spectrometric studies (Rinke and Schäfer, 1965; Büchler, Blackburn and Stauffer, 1966). It should be noted that the temperature of the electron diffraction experiment on Re_3Br_9 was between 300 and 400°C. The structure found in the vapour phase was in good agreement with the results of X-ray diffraction studies on crystalline rhenium halides (Cotton and Mague, 1964; Bennett, Cotton and Foxman, 1968). The Re_3X_9 units in the crystal are linked to each other by halogen bridges and the intermolecular bonds are of about the same length as the intramolecular bridging $Re-X$ bonds. This is illustrated in Fig. 34 for Re_3I_9.

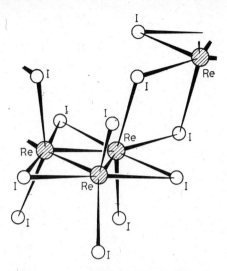

Fig. 34. Intra- and intermolecular bridging Re—I bonds in crystalline rhenium iodide (Bennett, Cotton and Foxman, 1968)

TETRAMERS

Both diethylaluminium fluoride, $(C_2H_5)_2AlF$ (Laubengayer and Lengnick, 1966), and dimethylaluminium fluoride, $(CH_3)_2AlF$ (Weidlein and Krieg, 1968), consist of tetrameric species in the liquid phase, according to spectroscopic studies. For the latter compound, the infrared and Raman spectra, containing only a few lines, were interpreted with a model of a planar eight-membered ring having D_{4h} symmetry. On the other hand, according to the electron diffraction results of Gundersen, Haugen and Haaland (1973), the tetrameric *dimethylaluminium fluoride*, $[(CH_3)_2AlF]_4$, molecules are non-planar in the vapour phase and even some high symmetry non-planar models (*e.g.*, C_{4v}, D_{2d}) could be ruled out. Excellent agreement was achieved with the experimental data for the model with C_s symmetry shown in Fig. 35. This may, however, be only one of the possible low symmetry models that would fit the measurements equally well. The following values were obtained for the most important bond distances and bond angles (in terms of r_a parameters):

123

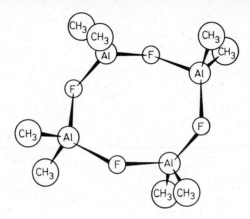

Fig. 35. Dimethylaluminium fluoride tetramer,
$[(CH_3)_2AlF]_4$

$$r(Al-C) = 1.947 \pm 0.004 \text{ Å}$$
$$r(Al-F) = 1.810 \pm 0.003 \text{ Å}$$
$$<C-Al-C = 131.2 \pm 1.9°$$
$$<F-Al-F = 92.3 \pm 1.2°$$
$$<Al-F-Al = 146.1 \pm 2.6°.$$

The Al−F bond is then considerably longer in this molecule than in monomeric AlF_3, while the Al−C bond is essentially of the same length as that of monomeric trimethylaluminium (see Table 7). Although the electron diffraction investigation did not establish the ring conformation unambiguously, it excluded with confidence the presence of appreciable amounts of trimers and dimers in the vapour phase. This finding is particularly important as the bond angles in the eight-membered rings are such that they could appear in a nearly planar six-membered ring also, imposing no strain. In this planar or nearly planar molecule, however, the Al−C bonds would be staggered with each other, producing an energetically very disadvantageous configuration. Accordingly, the formation of the stable tetrameric species may be accounted for by the properties of the rotational potential around the Al−F bond (Gundersen, Haugen and Haaland, 1973).

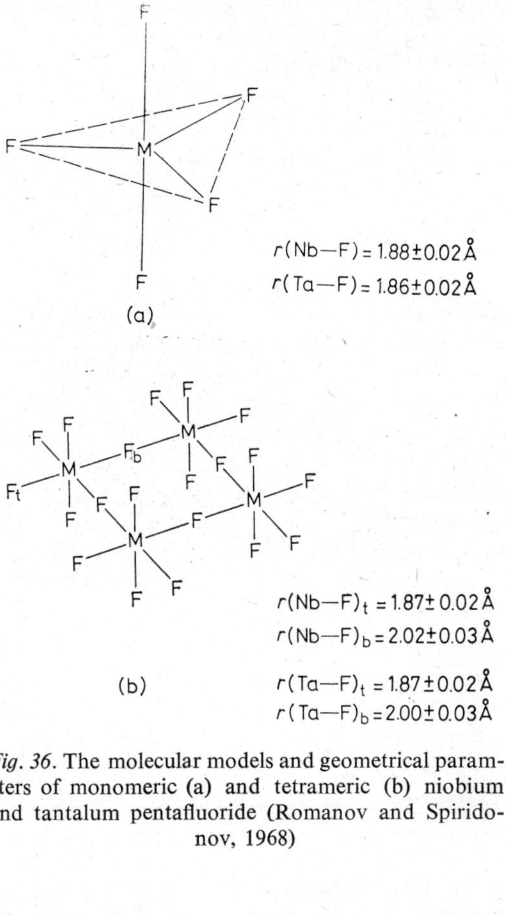

$r(\text{Nb}-\text{F}) = 1.88 \pm 0.02 \text{ Å}$
$r(\text{Ta}-\text{F}) = 1.86 \pm 0.02 \text{ Å}$

(a)

$r(\text{Nb}-\text{F})_t = 1.87 \pm 0.02 \text{ Å}$
$r(\text{Nb}-\text{F})_b = 2.02 \pm 0.03 \text{ Å}$

$r(\text{Ta}-\text{F})_t = 1.87 \pm 0.02 \text{ Å}$
$r(\text{Ta}-\text{F})_b = 2.00 \pm 0.03 \text{ Å}$

(b)

Fig. 36. The molecular models and geometrical parameters of monomeric (a) and tetrameric (b) niobium and tantalum pentafluoride (Romanov and Spiridonov, 1968)

The electron scattering pattern of *tantalum pentafluoride*, TaF$_5$, was observed to change considerably as the nozzle temperature varied. *Niobium pentafluoride*, NbF$_5$, showed similar behaviour. According to the findings of Romanov and Spiridonov (1968), the vapours of both compounds consist of tetrameric species with eight-membered rings at 30–50°C and monomeric species with a trigonal bipyramidal configuration at 200°C and above. The shape of the molecules and the bond distances determined are shown in Fig. 36. The terminal bonds in the two tetramers and also in the monomers are of the same length, within experimental error. The bridging bonds are considerably

125

longer. The structures of both tetramers in the crystal phase are similar to those in the vapour phase, having terminal bonds 0.1 Å shorter, according to the X-ray diffraction data by Edwards (1964). As the tetrameric species could not be detected in the mass spectra, it is of interest to note that recent electron diffraction experiments on tantalum pentafluoride provided additional evidence for the existence of tetrameric associates in the vapours (Spiridonov, 1974).

MIXED HALIDES

Numerous mixed halides with the general formula $M^nMe^mX_{n+m}$ (M and Me are n- and m-valent metals, X is a halogen) have recently been studied in the vapour phase by various physico-chemical techniques. Shol'ts and Sidorov (1972) suggested the following classification for the most common types of compounds, based on mass spectrometric studies and using the notation $A = M^IX$ and $B = Me^{II}X_2$ or $Me^{III}X_3$:

AB	$MMe^{II}X_3$,	for example	$NaBeF_3$
	$MMe^{III}X_4$,	for example	$NaAlF_4$
A_2B	$M_2Me^{II}X_4$,	for example	Na_2BeF_4
	$M_2Me^{III}X_5$,	for example	Na_2AlF_5
A_2B_2	$(MMe^{II}X_3)_2$,	for example	$(NaBeF_3)_2$
	$(MMe^{III}X_4)_2$,	for example	$(NaAlF_4)_2$

The vapour-phase molecular geometry has been determined for compounds belonging to the class AB only and these are dealt with below. However, based on analogy and spectroscopic evidence, Shol'ts and Sidorov (1972) advanced suggestions for the molecular configurations of the compounds of classes A_2B and A_2B_2 also and they are included in Fig. 37.

In the electron diffraction investigation of *lithium, sodium* and *potassium beryllium trifluoride*, Spiridonov, Erokhin and Brezgin (1972) first examined the molecular models with C_{2v} and C_{3v} symmetry shown in Fig. 38. Only partial results were obtained for $NaBeF_3$. It was suggested that the BeF_3 group has a regular triangular configuration with the beryllium atom in the centre. The length of the Be−F bond was found to be 1.49 Å. The findings for $LiBeF_3$ (Brezgin, 1972) and $KBeF_3$ (Spiridonov, Erokhin and Brezgin, 1972) were more complete. The BeF_3 groups in both molecules have the same configu-

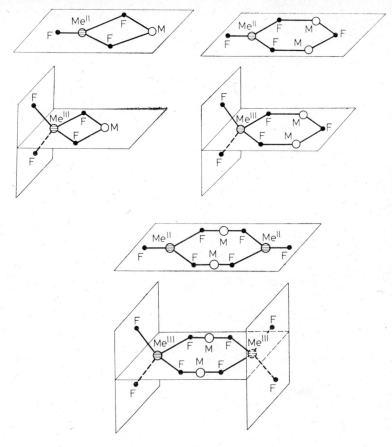

Fig. 37. Molecular configurations suggested for mixed halides by Shol'ts and Sidorov (1972)

ration as that in $NaBeF_3$, and $r(Be-F) = 1.49 \pm 0.01$ Å. Of those shown in Fig. 38, the best agreement with the experimental data was achieved for one of the models with C_{2v} symmetry. The lithium or potassium atom is located on the perpendicular to one of the edges of the triangle formed by the BeF_3 group, and this perpendicular passes through the beryllium atom (model I, Fig. 38). The $Li-F$ and $K-F$ bonds were found to be 1.72 ± 0.05 Å and 2.41 ± 0.02 Å, respectively.

127

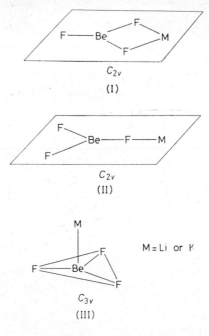

Fig. 38. Models of $M^I BeF_3$ molecules

It is interesting to compare the data on $r(Be-F)$ in gaseous and condensed phase mixed fluorides (Rahman, Fowler and Narten, 1972; Burns and Gordon, 1966) with those in BeF_2 found in the vapour phase (Akishin and Spiridonov, 1957) and condensed phase (Narten, 1972; Rahman, Fowler and Narten, 1972). The $Be-F$ bond length in the vapour-phase mixed fluorides is intermediate between those in BeF_2 gas (1.40 ± 0.03 Å) and condensed phase species (1.55–1.56 Å).

Various models have been tested against the experimental data in studies on $M^I Me^{III} X_4$ molecules. These models are illustrated in Fig. 39.

The spectra obtained in the first investigation on $LiAlF_4$ (Porter and Zeller, 1960; McCory, Paule and Margrave, 1963) were assigned to a model with C_{3v} symmetry ((III) in Fig. 39) in which three fluorine atoms are linked to the lithium atom and the latter is located on an axis perpendicular to one of the surfaces of the AlF_4 tetrahedron.

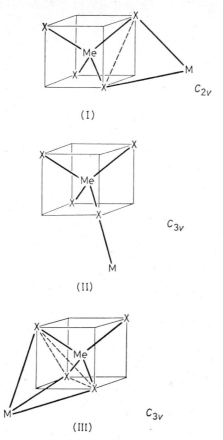

(I)

C_{2v}

(II)

C_{3v}

(III)

C_{3v}

Fig. 39. Models of $M^IMe^{III}X_4$ molecules

Büchler and Berkowitz-Mattuck (1967) gave preference to a model with C_{2v} symmetry ((I) in Fig. 39) by analogy with the bridge structures of Li_2X_2 and Al_2X_6 molecules. These models have been used by Sidorov and Kolosov (1968) in thermodynamic calculations. Matrix isolation infrared spectroscopic studies by Cyvin, Cyvin, Rao and Snelson (1971) and Cyvin, Cyvin and Snelson (1971) provided further evidence for the model with C_{2v} symmetry. However, the controversy is not considered to have been completely resolved.

An electron diffraction investigation of *sodium aluminium tetra-fluoride*, $NaAlF_4$, was carried out by Spiridonov and Erokhin (1969)

TABLE 26

Molecular parameters for $NaAlF_4$: internuclear distances (r) determined by electron diffraction and mean amplitudes of vibration (l) calculated from spectroscopic data

	r, Å*	l, Å**
$Al-F_t$	1.69 ± 0.02	0.058
$Al-F_b$	1.69 ± 0.02	0.063
$Na-F_b$	2.11 ± 0.02	0.139
$Al \dots Na$	2.58 ± 0.03	0.129
$F_b \dots F_b$	2.76 ± 0.03	0.113
$F_t \dots F_t$	2.76 ± 0.03	0.220
$F_b \dots F$	2.76 ± 0.03	0.152

* The temperature of the electron diffraction experiment was 1150 K, Spiridonov and Erokhin (1969).
** $T = 1000$ K, Cyvin, Cyvin and Snelson (1971).

and a structure containing two bridging fluorine atoms seems to be proved. The internuclear distances are given in Table 26. The good agreement between the experimental and theoretical distributions could be further improved by twisting the

triangle around the $F_b \dots F_b$ axis, as shown in Fig. 40, with an angle of about 26°. How much this effect can be considered to be real for the equilibrium structure is difficult to judge because of the experimental uncertainties and the large-amplitude motion, which is at least partly due to the high temperature of the experiment (1150 K).

The rich abundance of various species in the vapours of the mixed halides seems to be a typical feature. It is then extremely important to study carefully the composition of the vapour in parallel with the elucidation of the molecular geometry. In the electron diffraction investigation of $NaAlF_4$ the experimental conditions were chosen on the basis of mass spectrometric information (Sidorov, Erokhin,

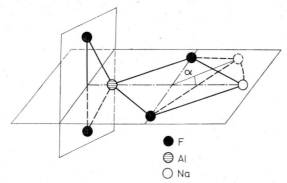

F

Al

Na

Fig. 40. The molecular configuration of $NaAlF_4$ (Spiridonov and Erokhin, 1969)

Akishin and Kolosov, 1967) in such a way as to ensure the maximum concentration of the monomeric species. Nevertheless, Spiridonov and Erokhin (1969) examined the possible effect of the presence of some proportion of dimeric species on the structural parameters determined. This effect turned out to be negligible owing to the large vibrational amplitudes associated with those distances which occur only in the dimeric molecules.

The results of an electron diffraction study on *potassium aluminium tetrachloride*, $KAlCl_4$, by Spiridonov, Erokhin and Lutoshkin (1971) showed the configuration of this molecule to be completely analogous to that of $NaAlF_4$. The deviation of the $Cl-Al-Cl$ bond angles from the tetrahedral value was found not to exceed $\pm 5°$. The terminal and bridging $Al-Cl$ bond distances are the same to within 0.1 Å. Other parameters are:

$$r(K-Cl) \quad = 2.84 \pm 0.02 \text{ Å}$$
$$r(Al-Cl) \quad = 2.16 \pm 0.02 \text{ Å}$$
$$<Cl-K-Cl = 77°.$$

A similar configuration was established for *potassium yttrium tetrachloride*, $KYCl_4$ (Spiridonov, Brezgin and Shakhparonov, 1971), with $r(K-Cl) = 2.85 \pm 0.01$ Å and $r(Y-Cl) = 2.54 \pm 0.01$ Å.

The electron diffraction investigation of *thallium indium tetrachloride*, $TlInCl_4$ (Spiridonov, Brezgin and Shakhparonov, 1972), provided a good opportunity for a more detailed elucidation of the molecular geometry than was the case for the above compounds. The reason is that the position of the "heavy" (large atomic number) thallium atom could be determined with more accuracy than those of the sodium or potassium atoms because of the larger scattering power. The overall geometry was found to be analogous to those of the other $MMe^{III}X_4$ halides. The geometrical parameters were found to be

$$r(In-Cl) = 2.37 \pm 0.01 \text{ Å}$$
$$r(Tl-Cl) = 2.91 \pm 0.02 \text{ Å}$$
$$<Cl-Tl-Cl = 83°.$$

The bond angles were determined to be tetrahedral to within 3° for the $InCl_4$ fragment and the $In-Cl$ bond lengths to be equal to within 0.1 Å. The twist around the $Cl_b \ldots Cl_b$ axis in the four-membered ring

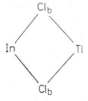

does not exceed 20° (see twist angle α in Fig. 40).

Some general observations can be made on the basis of the molecular geometries determined, following mostly the discussions by Szpiridonov (1972) and Rambidi (1973). The M^I-X bonds are 0.1–0.2 Å longer in all structures elucidated than the analogous bonds in the corresponding halide molecules. We refer to the data in Table 23 and for TlCl to the value of $r_e = 2.4848 \pm 0.0001$ Å determined by microwave spectroscopy (Mandel and Barrett, 1955). Also, the mean amplitudes of vibration associated with the M^I-X bonds are anomalously large, much larger than is usual for chemical bonds (0.03–0.07 Å). Thus, e.g., the values $l(K-Cl) = 0.13$ Å in $KAlCl_4$ (Szpiridonov, Erokhin and Lutoshkin, 1971) and $l(Na-F) = 0.14$ Å in $NaAlF_4$ (Cyvin, Cyvin and Snelson, 1971) indicate a rather weak alkali metal–halogen bond and a relatively loose linkage between the metal atom and the remainder of the molecule.

The AlX_4 and the other tetrahedral fragments, on the other hand, appear to be largely autonomous and relatively rigid. This is indicated by the regular or nearly regular geometrical configuration, equal or nearly equal bond distances and, at least for $NaAlF_4$, by the nearly equal mean amplitudes of vibration, $l(Me-X)_t$ and $l(Me-X)_b$ given in Table 26. Similar conclusions probably apply to the beryllium trifluoride moieties in $LiBeF_3$, $NaBeF_3$ and $KBeF_3$.

Considering the behaviour of the MeX_4 fragments, it is interesting to note that the AlX_4 groups of the $MAlX_4$ systems display similar structural properties in the condensed phase (cf., Szpiridonov, 1972). This in turn suggests that the transition from one phase into the other does not influence appreciably the structure of this fragment.

There is a considerable difference between the structure of the aluminium tetrachloride fragment of the mixed halides and the configuration around the aluminium atom in aluminium trichloride dimer. The terminal $Al-Cl$ bond is significantly shorter than the bridging $Al-Cl$ bond in Al_2Cl_6 (Table 24). The length of the $Al-Cl$ bond of the regular or nearly regular tetrahedral $AlCl_4$ fragment of the mixed halides appears to be intermediate between the lengths of the two types of $Al-Cl$ bonds in the strongly distorted tetrahedral $AlCl_4$ group of the Al_2Cl_6 molecule.

To summarize, there is great mobility of the M metal atom on one side, and symmetrical configuration and rigidity of the AlX_4 fragment on the other side in the $MAlX_4$ structures. These properties indicate a dynamic configuration rather than a static geometry. In the dynamic establishment, the M atom belongs to all four halogen atoms with equal probability. Accordingly, the structural formula $M^{\delta+}[AlX_4]^{\delta-}$, which also expresses the polar character of the linkage, was suggested by Szpiridonov (1972).

133

SALTS OF OXYACIDS

Much experimental evidence has accumulated on the existence of stable molecules in the vapours of salts of numerous oxyacids. Studies on the processes of evaporation by mass spectrometry and other physico-chemical techniques were performed, among others, on metaborates (Hildenbrand, Hall and Potter, 1963), alkali metal nitrates (Gordon and Campbell, 1955; Hardy and Field, 1963), copper(II) nitrate (Porter, Schoomaker and Addison, 1959), caesium sulphate (Spitsyn and Shostak, 1949), alkali metal chromates (Afonskii, 1962), indium molybdate (Berns, De Maria, Drowart and Inghram, 1963) and thallium perrhenate (Semenov, 1970). High-temperature electron diffraction is used almost exclusively for determining the molecular geometries of these low-volatility compounds, and it is very useful to supplement these studies with spectroscopic information. The matrix isolation technique has special importance for this class of compounds.

The experimental and interpretational difficulties connected with the application of electron diffraction at high temperatures have already been mentioned. As the technique improves, it is desirable to repeat certain investigations and some of the salts of oxyacids provide ample proof of how important such reinvestigations may be. Repeated studies may be aimed at testing the validity of fundamental conclusions on the shapes of the molecules and also at gaining more and more insight into the details of structural peculiarities. It was the reinvestigation of the molecular geometries of Me_2XO_4 salts (where Me = monovalent metal and X = S, Cr, Mo or W) that demonstrated the coordination character of these structures. It is not so obvious that the metaborates and nitrates are also coordination compounds, but they can be considered at least to be borderline cases. The molecular geometries of the following salts of oxyacids have been studied so far:

$LiBO_2$	$LiNO_3$	$KReO_4$	K_2SO_4	Na_2CrO_4	Cs_2MoO_4	Cs_2WO_4
$NaBO_2$	$NaNO_3$	$TlReO_4$	Cs_2SO_4	K_2CrO_4	In_2MoO_4	$BaWO_4$
KBO_2	$Cu(NO_3)_2$		Tl_2SO_4		Tl_2MoO_4	
$RbBO_2$						
$CsBO_2$						
$TlBO_2$						

ALKALI METAL METABORATES

We start the discussion of the results on the alkali metal metaborates by mentioning the studies on *boron oxide*, B_2O_3, and *metaboric acid*, HBO_2. Numerous models have been tested in electron diffraction and spectroscopic studies on boron oxide, only a few of which are presented here:

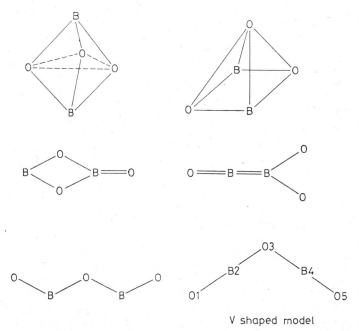

V shaped model

The visual electron diffraction study by Akishin and Spiridonov (1960) established the V-shaped model, in agreement with the findings of a vibrational spectroscopic study (White, Walsh and Mann, 1958). The following parameters have also been determined: $r(B-O) =$

= 1.36 ± 0.02 Å, $r(B = O) = 1.20 ± 0.03$ Å and B−O−B bond angle = 95°. The O−B−O chain was found to be linear. This observation was subsequently supplemented by further electron diffraction evidence and more accurate parameters were eventually obtained (Ezhov, Tolmachev, Spiridonov and Rambidi, 1968; Ezhov, Tolmachev and Rambidi, 1970). The electron scattering pattern is primarily determined by the relatively rigid O−B−O fragments, while the information on the B−O−B bond angle is rather uncertain as a consequence of the large-amplitude motion. The question of whether the two types of boron–oxygen bonds can or cannot be distinguished on the basis of the electron diffraction data was treated very cautiously in the papers cited above. Based on careful considerations, it was concluded that the difference between the two bond distances does not exceed 0.1 Å. An infrared spectroscopic study in an argon matrix (Sommer, White, Linevsky and Mann, 1963) verified the V-shaped molecular configuration with C_{2v} symmetry but disagreed with the early electron diffraction result concerning the small value of the B−O−B bond angle. The most recent electron diffraction investigation utilized intensity data in the important small scattering angle region and yielded the value of 132 ± 5° for the B−O−B bond angle. The following internuclear distances and mean amplitudes of vibration were obtained:

	r_g, Å	l, Å
B−O (average)	$1.26_5 ± 0.01$	$0.11 ± 0.01$
O1...O3	$2.53 ± 0.01$	$0.11 ± 0.01$
B2...O5	$3.52 ± 0.02$	$0.14 ± 0.02$
B2...B4	$2.33 ± 0.05$	$0.20 ± 0.05$
O1...O5	$4.66 ± 0.10$	$0.35 ± 0.10$

The structures of the alkali metal metaborates closely resemble the geometry of boron oxide in that they also contain a linear O−B=O chain. The shape of their molecules is, in fact, the same as that of metaboric acid:

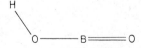

which was established by a gas-phase infrared spectroscopic investigation (White, Mann, Walsh and Sommer, 1960).

There is good agreement between the findings of the infrared spectroscopic (Büchler and Marram, 1963; Seshadri, Nimon and White, 1969) and electron diffraction (Akishin and Spiridonov, 1961a, 1962; Ezhov, Tolmachev, Spiridonov and Rambidi, 1968; Tolmachev, 1970; Ezhov, Tolmachev and Rambidi, 1972; Ezhov, Komarov and Tolmachev, 1973) studies concerning the molecular geometries of alkali metal metaborates and thallium(I) metaborate. Painstaking reinvestigations have gradually improved the data, as in the case of boron oxide, and the most recent results are summarized in Table 27, corresponding to experiments at the following temperatures:

$LiBO_2$ 1550 ± 50 K
$NaBO_2$ 1480 ± 50 K
KBO_2 1280 ± 50 K
$RbBO_2$ 1200 ± 50 K
$CsBO_2$ 1100 ± 50 K
$TlBO_2$ 1100 ± 50 K.

The uncertainties indicated in Table 27 for the parameters $r(B=O)$ and $r(B-O)$ require some clarification. The electron diffraction analysis, in fact, produced an average value for the bond distances. It was also determined that the difference between the lengths of the two types of bond distances does not exceed 0.1 Å. Thus, the bond distances presented in Table 27 refer to the extreme values possible with the corresponding error limits (Ezhov, Tolmachev and Rambidi, 1972).

The $Me-O$ bonds in the metaborates appear to be longer than the analogous bonds in the related oxides and hydroxides. This may be an indication of a weaker metal–oxygen linkage in the metaborates. There are descriptions of the bonding properties in the literature that emphasize either the ionic or the covalent character of the bonding in the metaborates (see, for example, Büchler and Marram, 1963; Seshadri, Nimon and White, 1969). According to Spiridonov (1974), a description with a dynamic model similar to those of mixed halides or the salts of oxyacids (p. 152) may be applicable to metaborates as well.

TABLE 27

Bond distances and bond angles in alkali metal metaborates and some related simple compounds

Molecule		$r(B=O)$, Å	$r(B-O)$, Å	$<O-B-O$	$r(Me-O)$, Å	$<Me-O-B$
B_2O_3	(a)	$1.21 \begin{smallmatrix}+0.05\\-0.01\end{smallmatrix}$	$1.31 \begin{smallmatrix}+0.01\\-0.05\end{smallmatrix}$	$180°$		
$LiBO_2$	(b)	$1.19 \begin{smallmatrix}+0.05\\-0.01\end{smallmatrix}$	$1.29 \begin{smallmatrix}+0.01\\-0.05\end{smallmatrix}$	$180°$		
Li_2O	(c)				1.60 ± 0.02	
$NaBO_2$	(b)	$1.20 \begin{smallmatrix}+0.05\\-0.01\end{smallmatrix}$	$1.30 \begin{smallmatrix}+0.01\\-0.05\end{smallmatrix}$	$180°$	2.14 ± 0.03	$106 \pm 5°$
KBO_2	(b)	$1.21 \begin{smallmatrix}+0.05\\-0.01\end{smallmatrix}$	$1.31 \begin{smallmatrix}+0.01\\-0.05\end{smallmatrix}$	$180°$	2.36 ± 0.03	$100 \pm 5°$
KOH	(d)				2.18 ± 0.01	
$RbBO_2$	(b)	$1.21 \begin{smallmatrix}+0.05\\-0.01\end{smallmatrix}$	$1.31 \begin{smallmatrix}+0.01\\-0.05\end{smallmatrix}$	$180°$	2.54 ± 0.02	
$RbOH$	(e)				2.301 ± 0.002	
$CsBO_2$	(f)	1.25 ± 0.01 (mean value)		$180°$	2.65 ± 0.05	$140 \pm 5°$
$CsOH$	(e)				2.391 ± 0.002	
	(g)				2.40 ± 0.01	
$TlBO_2$	(f)	1.25 ± 0.01 (mean value)		$180°$	2.38 ± 0.05	$140 \pm 5°$
Tl_2O	(a)				2.15 ± 0.02	

(a), (b), (c), (f), (g) Electron diffraction.

(d), (e) Microwave spectroscopy.

(a) Tolmachev (1970), the values of uncertainties were obtained as described for metaborates (see p. 137).

(b) r_g parameters, Ezhov, Tolmachev and Rambidi (1972).

(c) Tolmachev, Zasorin and Rambidi (1969).

(d) Kuczkowski, Lide and Krisher (1966).

(e) Equilibrium bond distance (r_e), Lide and Matsumura 1969).

(f) Ezhov, Komarov and Tolmachev (1973).

(g) Ugarov, Tolmatchev, Ezhov and Rambidi (1972).

NITRATES

The importance of our knowledge about the structure of *nitric acid,* HNO_3, for discussing the results on nitrates is obvious. According to the microwave spectroscopic studies of Millen and Morton (1960) and Cox and Riveros (1965), the HNO$_3$ molecule is planar and the NO$_2$ group is slightly tilted from the O−H bond:

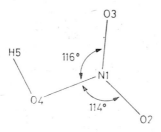

The geometrical parameters determined by microwave spectroscopy are presented in Table 28, together with the results of a less complete visual electron diffraction investigation by Akishin, Vilkov and Rosolovskii (1960). The general agreement between the data produced by the two techniques is satisfactory, considering the experimental error.

For *lithium nitrate*, LiNO$_3$, and *sodium nitrate*, NaNO$_3$, molecular structures analogous to that of nitric acid were elucidated by sector-microphotometer electron diffraction (Khodchenkov, Spiridonov and

TABLE 28

*Molecular geometry of vapour-phase nitric acid**

	(a)	(b)	(c)
r(N1−O2), Å	1.206±0.005	1.199	1.22±0.01
r(N1−O3), Å		1.211	
r(N1−O4), Å	1.405±0.005	1.406	1.40±0.01
r(O−H), Å	0.96 (assumed)	0.964	
<O2−N1−O3	130° ±20′	130°16′	135 ±2.5°
<O2−N1−O4	114° ±20′	113°51′	
<O3−N1−O4	116° ±20′	115°53′	
<N1−O4−H5	102° ±30′	102°9′	

* The numbering of atoms is given above.
(a) and (b) Microwave spectroscopy, (a) Millen and Morton (1960).
(b) Cox and Riveros (1965).
(c) Visual technique of electron diffraction, Akishin, Vilkov and Rosolovskii (1960).

139

Akishin, 1965). The following geometrical parameters were obtained:

	$LiNO_3$	$NaNO_3$
$r(Me-O)$, Å	1.60	1.90
$r(N-O)$, Å	1.40	1.40
$r(N=O)$, Å	1.22	1.22
$<Me-O-N$	105°	105°
$<O=N=O$	135°	135°.

The uncertainty in the bond distances was given as 1%. The validity of the above model for the nitrate molecules is strongly questioned in the light of more recent data on $TlNO_3$ (Ishchenko, Spiridonov and Zasorin, 1974). Accordingly, a reinvestigation seems to be worth while.

Copper(II) nitrate, $Cu(NO_3)_2$, is a surprisingly volatile compound (Addison and Hathaway, 1958). A nozzle temperature of 175–185°C was sufficient in the first electron scattering experiments that were performed without a rotating sector (LaVilla and Bauer, 1963). The elucidation of the molecular configuration proved to be difficult as the theoretical distributions calculated for several models showed good agreement with the measured data. An earlier study (Bauer and Addison, 1960) concluded, erroneously as it was shown later, that the two nitrate groups are linked to the copper atom differently. According to the more recent data, the structure has higher symmetry. Of the models considered, three are presented below:

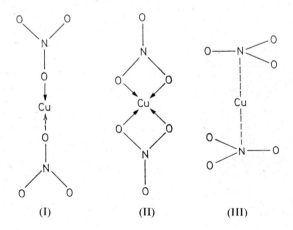

(I)   (II)   (III)

The possibility of a twist around the N—Cu—N axis was examined in detail for model (II). As the final conclusion, a planar model (II) was given by the authors. A similar result was reported later by Khodchenkov (1965) based on an independent electron diffraction investigation. The geometrical parameters obtained in the two studies are also in good agreement and they include:

$$r(Cu-O) = 2.00 \pm 0.02 \text{ Å}$$
$$r(Cu-N) = 2.30 \pm 0.03 \text{ Å}$$
$$<O-N-O = 120 \pm 2°$$
$$<O-Cu-O = 70°.$$

Khodchenkov (1965) indicated only an average value (1.30 Å) for the nitrogen–oxygen bond distances although he had been using two different values in testing the models. Incidentally, these values were the same as those reported by LaVilla and Bauer (1963):

$$r(N=O) = 1.24 \text{ Å}$$
$$r(N-O) = 1.32 \text{ Å}.$$

A more recent electron diffraction study by Ishchenko, Zasorin, Spiridonov, Bersuker and Budnikov (1974) verified model (II) for $Cu(NO_3)_2$ with each $CuNO_3$ group

being planar. The two nitrate groups, however, were not found to be coplanar. Molecular orbital calculations predicted a dihedral angle $\Theta = 90°$ in the equilibrium configuration. It was also estimated that an average angle of 84.5° would be obtained from the electron diffraction data at the temperature of the experiment (190 $\pm$ 5°C). It was not reported whether this calculated value has been tested against the measured data. The geometrical parameters determined in this electron diffraction study are as follows:

$$r_a(Cu-O) = 1.96 \pm 0.02 \text{ Å}$$
$$r_a(N-O) = 1.31 \pm 0.01 \text{ Å}$$
$$r_a(N=O) = 1.23 \pm 0.01 \text{ Å}$$
$$<O-Cu-O = 70 \pm 2°.$$

The length of the N=O bond was found to be the same in both the vapour and the crystal (1.218 ± 0.004 Å) of copper(II) nitrate as the latter was determined by X-ray diffraction (Sass, Vidale and Donohue, 1957). On the other hand, the crystalline molecular configuration is different from that of the vapour-phase molecules. There are parallel chains in the crystal consisting of NO_3 groups linked to each other by Cu atoms (Wallwork, 1959):

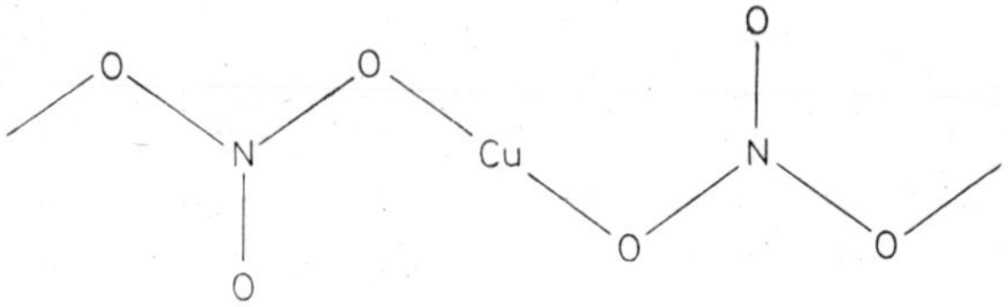

It was suggested (Addison and Hathaway, 1958) that these chains are broken up into molecules during sublimation.

The preliminary results of the electron diffraction study on thallium(I) nitrate, $TlNO_3$, indicated (Ishchenko, Spiridonov and Zasorin, 1974) the presence of a dynamical structure in the vapour phase. The experimental data could be best approximated by using the following three models with relative abundances of 18, 71 and 11% of models (I), (II) and (III), respectively:

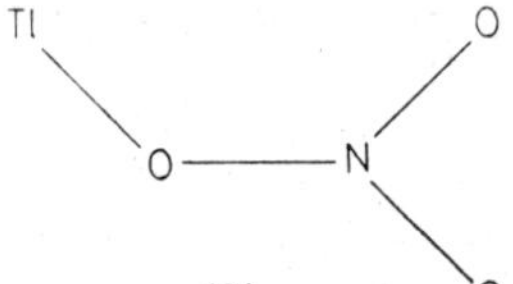

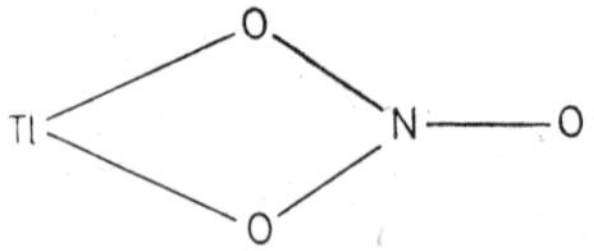

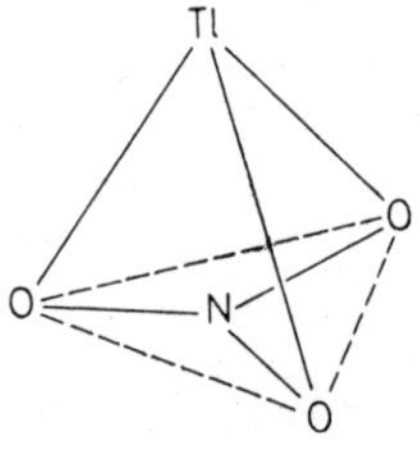

The structures of alkali metal nitrates, $TlNO_3$ and $Cu(NO_3)_2$ isolated in matrices have been studied by Raman and infrared spectroscopy (Smith, James and Devlin, 1971). It was established that in all instances the spectra originated primarily from stable monomeric species. Only for the alkali metal nitrates could any indication of the presence of a certain proportion of dimeric species be found. For the latter, the following structure was suggested:

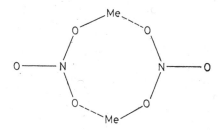

Considerable distortion of the nitrate group was observed in the alkali metal nitrate monomers, which was ascribed to the polarizing effect of the cation. The distortion was found to be much less in $TlNO_3$ and $Cu(NO_3)_2$, which both have more covalent character. The infrared spectrum of copper(II) nitrate supported the model with D_{2h} symmetry suggested by LaVilla and Bauer (1963).

An electron diffraction structural analysis on the molecular geometry of *zirconium tetranitrate*, $Zr(NO_3)_4$, was performed by Tusccv, Zasorin and Spiridonov (1974). The planarity of the

fragment was established. The zirconium–nitrogen internuclear distance was found to be 2.650 ± 0.014 Å with an l value of 0.08 ± 0.02 Å. The following bond distances and vibrational amplitudes were also determined:

	r_a, Å	l, Å
Zr–O	2.219 ± 0.006	0.052 ± 0.012
N–O	1.284 ± 0.007	0.038 ± 0.016
N=O	1.184 ± 0.010	0.036 ± 0.017

143

The uncertainties indicated are not total errors but are the standard deviations obtained in the least-squares fit. The configuration of the ligands around the zirconium atom was found to be distorted tetrahedral.

MeReO$_4$ (Me = K or Tl), Me$_2$XO$_4$ (Me = Na, K, Cs, In or Tl and X = S, Cr, Mo or W) and BaWO$_4$ MOLECULES

The electron diffraction experiments for most of the compounds in this class required particularly high temperatures, as can be seen from the following nozzle temperatures:

KReO$_4$	800– 900°C
TlReO$_4$	570°C
K$_2$SO$_4$	1300–1500°C
Tl$_2$SO$_4$	700°C
Na$_2$CrO$_4$	1000–1200°C
Cs$_2$MoO$_4$	1050°C
In$_2$MoO$_4$	1000°C
Tl$_2$MoO$_4$	630– 730°C
Cs$_2$WO$_4$	900°C
BaWO$_4$	1800–2000°C.

The visual electron diffraction data on KReO$_4$ (Spiridonov, Khodchenkov and Akishin, 1965a), Cs$_2$SO$_4$ (Spiridonov, Khodchenkov

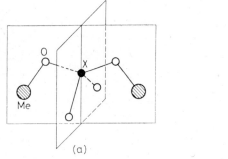 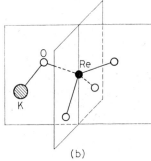

(a)   (b)

Fig. 41. Previously assumed models for the gaseous Me$_2$XO$_4$ (Me = K, Cs or Na; X = S or Cr) and KReO$_4$ molecules

and Akishin, 1965b), Na_2CrO_4 and K_2CrO_4 (Spiridonov, Khodchen-kov and Akishin, 1965c) were interpreted in terms of the classical covalent structures. According to this model, the geometrical arrange-ment is that of a distorted tetrahedron with the X atom at the centre and the oxygen atoms at the apexes. The alkali metal atom is linked to the oxygen atom as indicated in Fig. 41 (a) and (b). There are two different types of X – O bonds in such a structure, roughly correspond-ing to single and double bonds as illustrated in the classical structural formulae

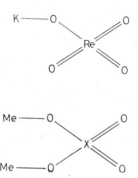

The parameters obtained by the visual technique are given in Table 29.

The structural pattern described above, which is so familiar from elementary textbooks, differed considerably from that established for this type of compounds in solid and liquid states. For the latter, a regular or nearly regular tetrahedral arrangement of the oxygen atoms around the central atom was found (see, for example, the data in Table 33 and the references therein).

An important development in the structural chemistry of the salts of oxyacids occurred when Büchler, Stauffer and Klemperer (1967) reported the absence of a permanent electric dipole moment of the Cs_2SO_4 molecules as determined by molecular beam deflection measurements. As this finding was contradictory to the structure elucidated from the visual electron diffraction study, Spiridonov and Lutoshkin (1970) repeated the investigation using the sector-micro-photometer technique. In addition, they also examined the structure of *potassium sulphate*, K_2SO_4, and *potassium chromate*, K_2CrO_4. Other modern electron diffraction studies have since been performed on *caesium molybdate*, Cs_2MoO_4, and *caesium tungstate*, Cs_2WO_4, by

10

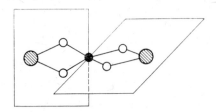

Fig. 42. The "bicyclic structure" of salts of oxyacids containing XO_4 (X = S, Cr, Mo or W) group

Ugarov, Ezhov and Rambidi (1973a), *thallium perrhenate,* $TlReO_4$, by Roddatis, Tolmachev, Ugarov, Ezhov and Rambidi (1974), and *barium tungstate,* $BaWO_4$, by Ivanov, Spiridonov, Erokhin and Levitskii (1973). The molecular geometries determined from the sector-micro-photometer studies are markedly different from those derived from the visual electron diffraction data. Accordingly, the other compounds not yet reinvestigated require a new study.

As for the structures obtained using up-to-date techniques, the XO_4 groups were found to be regular tetrahedral within the experimental error. The alkali metal (or Tl or Ba) atoms are located on axes perpendicular to the edges of the XO_4 tetrahedron, as indicated in Fig. 42. The name "bicyclic structure" is also used for this model of Me_2XO_4, as it consists of two

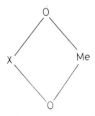

four-membered rings whose planes are perpendicular to each other.

The molecular parameters are given in Tables 30 and 31. The values obtained for $r(S-O)$ are considerably smaller than that which can be considered to be the single bond (1.69 Å).* As values of about

---

* This was calculated on the basis of the equation of Schomaker and Stevenson (1941) using the covalent radii given by Pauling (1960).

TABLE 29

Geometrical parameters* for compounds KReO$_4$ and
Me$_2$XO$_4$ (Me = Cs, Na or K and X = S or Cr)
determined by the visual technique of electron diffraction

Parameters	KReO$_4$ (a)	Cs$_2$SO$_4$ (b)	Na$_2$CrO$_4$ (c)	K$_2$CrO$_4$ (c)
r(X=O), Å	1.75	1.43	1.60	1.60
r(X−O), Å	1.95	1.62	1.80	1.80
r(Me−O), Å	2.20	2.50	1.90	2.20
<O=X=O		120°	135°	135°
<O−X−O		105°	105°	105°
<O−X=O	95°			
<Me−O−X	105°	105°	105°	105°

* The uncertainties of the parameters are not given in the papers cited.
(a) Spiridonov, Khodchenkov and Akishin (1965a).
(b) Spiridonov, Khodchenkov and Akishin (1965b).
(c) Spiridonov, Khodchenkov and Akishin (1965c).
Note: The results of an electron diffraction reinvestigation using the sector-microphotometer technique for Cs$_2$SO$_4$ and K$_2$SO$_4$ are given in Table 30. These results differ considerably from those presented above (see the text for more detail). Accordingly, a reinvestigation for the other compounds is felt to be needed.

TABLE 30

Molecular parameters for K$_2$SO$_4$, Cs$_2$SO$_4$ and K$_2$CrO$_4$ determined
by Spiridonov and Lutoshkin (1970) using the sector technique
of electron diffraction*

Parameters	K$_2$SO$_4$	Cs$_2$SO$_4$	K$_2$CrO$_4$
r(X−O), Å	1.47±0.01	1.48±0.01	1.66±0.01
r(Me−O), Å	2.45±0.03	2.60±0.03	2.45±0.03
<O−Me−O	59°**	55°**	67°**
l(Me−O), Å	0.15**	0.25**	0.12**

* The XO$_4$ group is regular tetrahedral with an uncertainty of ±10° for the bond angles, O−X−O; the possible difference between the lengths of the bonds X−O is not larger than 0.1 Å, as noted by Spiridonov and Lutoshkin (1970).
** The uncertainties of these parameters are not given in the paper cited.

TABLE 31

Molecular parameters for Cs_2MoO_4 and Cs_2WO_4 determined by Ugarov, Ezhov and Rambidi (1973a) using the sector technique of electron diffraction

		Cs_2MoO_4	Cs_2WO_4
$r(X-O)$,	Å	1.80 ± 0.02	1.82 ± 0.02
$r(Cs-O)$,	Å	2.80 ± 0.03	2.78 ± 0.03
$r(Cs...X)$,	Å	3.50 ± 0.03	3.50 ± 0.03
$r(Cs...O)$,	Å	4.81 ± 0.10	4.84 ± 0.10
$r(Cs...Cs)$,	Å	7.00 (calculated)	7.00 ± 0.12
$<O-X-O$		$105 \pm 4°$	$104 \pm 4°$
$l(X-O)$,	Å	0.06 ± 0.02	0.06 ± 0.02
$l(Cs-O)$,	Å	0.26 ± 0.03	0.22 ± 0.03
$l(Cs...X)$,	Å	0.21 ± 0.02	0.18 ± 0.02
$l(Cs...O)$,	Å	0.45 ± 0.05	0.35 ± 0.04
$l(Cs...Cs)$,	Å	0.50	0.45 ± 0.05

TABLE 32

Comparison of the bond distances in Me_2XO_4 salts ($Me = K$ or Cs and $X = S, Cr, Mo$ or W) and some related simple compounds

Molecule		Bond	Distance, Å
LiF	(a1)	Li−F	1.563892 ± 0.000050
Li_2O	(b1)	Li−O	1.60 ± 0.02
KF	(a1)	K−F	2.17144 ± 0.00005
KOH	(a2)	K−O	2.18 ± 0.01
K_2SO_4	(b2)	K−O	2.45 ± 0.03
K_2CrO_4	(b2)	K−O	2.45 ± 0.03
CsF	(a1)	Cs−F	2.3453 ± 0.0001
CsOH	(a3)	Cs−O	2.391 ± 0.002
	(b3)	Cs−O	2.40 ± 0.01
Cs_2SO_4	(b2)	Cs−O	2.60 ± 0.03
Cs_2MoO_4	(b4)	Cs−O	2.80 ± 0.03
Cs_2WO_4	(b4)	Cs−O	2.78 ± 0.03

(a) Microwave spectroscopy, (1) for references, see Lide, Cahill and Gold (1964); (2) Kuczkowski, Lide and Krisher (1966); (3) Lide and Matsumura (1969).

(b) Sector technique of electron diffraction, (1) Tolmachev, Zasorin and Rambidi (1969); (2) Spiridonov and Lutoshkin (1970); (3) Ugarov, Tolmachev, Ezhov and Rambidi (1972); (4) r_g parameters, Ugarov, Ezhov and Rambidi (1973a).

1.40–1.44 Å are usually assigned to the S−O double bond (Banister, Moore and Padley, 1968), the sulphur–oxygen linkage in K_2SO_4 and Cs_2SO_4 in the gas phase can be considered to be of double bond character.

According to recent results, it can be concluded that the structures of the salts in the vapour phase and those of the corresponding acids are different. The molecular geometry of sulphuric acid has so far been elucidated in the crystalline phase only (Claudine, 1965). Two different sulphur–oxygen bonds were observed by X-ray diffraction:

$$r(S-O) = 1.535 \pm 0.015 \text{ Å}$$
$$r(S=O) = 1.426 \pm 0.015 \text{ Å}.$$

Similarly, the vapour-phase structures of nitric acid (Millen and Morton, 1960; Cox and Riveros, 1965; Akishin, Vilkov and Rosolovskii, 1960) and perchloric acid (Akishin, Vilkov and Rosolovskii, 1959; Clark, Beagley and Cruickshank, 1968) are characterized by two different central atom–oxygen bonds.

It is also interesting to compare the structures of the chromate, molybdate and tungstate groups with those of polymeric oxides (cf., p. 166). The lengths of the X−O bonds (X = Cr or Mo) in K_2CrO_4 and Cs_2MoO_4 are intermediate between those of the bridging and terminal bonds in the oxides.

The lengths of the alkali atom–oxygen bonds in the salts of oxyacids are compared with analogous data for other compounds in Table 32. The Me−O bonds in the salts of oxyacids appear to be longer than those in the hydroxides, which in turn have lengths similar to the Me−F bonds in corresponding fluorides, as expected.

The structural peculiarities of the salts of oxyacids deserve attention. One of their interesting features is the particularly large mean amplitude of vibration associated with the Me−O bonds. Some of these amplitudes are given in Tables 30 and 31 and others are mentioned below. Both the lengths and the vibrational amplitudes of the Me−O bonds indicate that the interaction between the Me atom and the remainder of the molecule is rather weak.

Comparison of the structures of the salts in different phases is facilitated by the fact that numerous X-ray diffraction studies have been performed on crystalline compounds containing SO_4^{2-}, CrO_4^{2-}, MoO_4^{2-} or WO_4^{2-} ions (see, for example, Sutton, 1958, 1965; Banister,

TABLE 33

Comparison of $r(X-O)$ bond distances of the tetrahedral XO_4 groups in the molecules of salts in gaseous, liquid and solid states

Compounds	Bond distances, Å	Phases	Techniques*	References
Sulphates	$r(S-O)$			
Li_2SO_4	1.46	Liquid	XD	(a)
Na_2SO_4	1.48	Liquid	XD	(a)
K_2SO_4	1.46	Liquid	XD	(a)
	1.47 ± 0.01	Gaseous	ED	(b)
Cs_2SO_4	1.48 ± 0.01	Gaseous	ED	(b)
Tl_2SO_4	1.48 ± 0.01	Gaseous	ED	(c)
Chromates	$r(Cr-O)$			
K_2CrO_4	1.66 ± 0.01	Gaseous	ED	(b)
$PbCrO_4$	1.65	Solid	XD	(d)
Molybdates	$r(Mo-O)$			
K_2MoO_4	1.76 ± 0.01	Solid	XD	(e)
Cs_2MoO_4	1.80 ± 0.01	Gaseous	ED	(f)
In_2MoO_4	1.81 ± 0.02	Gaseous	ED	(g)
Tl_2MoO_4	1.81 ± 0.02	Gaseous	ED	(h)
$MnMoO_4$	1.761	Solid	XD	(i)

* XD = X-ray diffraction, ED = Electron diffraction.
(a) Ukshe (1963).
(b) Spiridonov and Lutoshkin (1970).
(c) Ugarov, Ezhov and Rambidi (1973b).
(d) Sutton (1965).
(e) Gatehouse and Leverett (1969).
(f) Ugarov, Ezhov and Rambidi (1973a).
(g) Tolmachev and Rambidi (1971).
(h) Tolmachev and Rambidi (1972).
(i) Abrahams and Reddy (1965).

TABLE 33 (continued)

Compounds	Bond distances, Å	Phases	Techniques*	References
Tungstates	$r(W-O)$			
Li_2WO_4	1.79	Solid	XD	(j)
K_2WO_4	1.79	Solid	XD	(k)
Cs_2WO_4	1.82	Gaseous	ED	(f)
$CaWO_4$	1.785	Solid	XD	(l)
	1.788	Solid	XD	(m)
	1.771	Solid	XD	(n)
$Al_2(WO_4)_3$	1.783 ± 0.023	Solid	XD	(o)
$Sc_2(WO_4)_3$	1.761 ± 0.046	Solid	XD	(p)

(j) Zachariasen and Plettinger (1961).
(k) Koster, Kools and Rieck (1969).
(l) Kay, Frazer and Almodovar (1964).
(m) Zalkin and Templeton (1964).
(n) Burbank (1965).
(o) Craig and Stephenson (1968).
(p) Abrahams and Bernstein (1966).

Moore and Padley, 1968; and the references in Table 33). The XO_4^{2-} ion appears to have a tetrahedral configuration in the crystal. Regular tetrahedra of SO_4^{2-} ions were also observed by X-ray diffraction in melted samples of Li_2SO_4, Na_2SO_4 and K_2SO_4 (Ukshe, 1966). As far as the lengths of the $S-O$ bonds are concerned, the crystal-phase data are too scattered for a meaningful comparison to be made. Some more recent and accurate data were collected in a review by Banister, Moore and Padley (1968) on hydrated sulphates. The value of $r(S-O)$ is 1.473 ± 0.003 Å, in good agreement with the vapour-phase results. Similarly, the geometrical parameters for Li_2SO_4, Na_2SO_4 and K_2SO_4 shown in Table 33 and referring to the vapour and liquid phases do not appear to be considerably different. The same can be said about the data on molybdates and tungstates. Inspection of the data indicates that the structure of the XO_4 groups is little influenced by the phase or even by the nature of the alkali metal atom, Me.

In the molybdates and tungstates of alkali metals, the central atom is always surrounded by four oxygen atoms in a tetrahedral arrangement, while in the case of other metals octahedral XO_6 structures have also been found. Among both the tetrahedral and octahedral systems a constancy of the $r(X-O)$ values can be observed but there are considerable differences between those referring to the two groups. To illustrate this, some $X-O$ distances for octahedral structures are given below and can be compared with the values for tetrahedral arrangements given in Table 33:

Compound $MeXO_4$	$r(X-O)$, Å (average value)	
$CoMoO_4$	1.991	cited in Abrahams and Reddy (1965)
$NiWO_4$	1.92 ± 0.21	Keeling (1957)
$MnWO_4$	1.95 ± 0.19	cited in Abrahams and Bernstein (1966)
$CuWO_4$	1.941	Kihlborg and Gebert (1970)
$FeWO_4$	1.940	cited in Kihlborg and Gebert (1970)

Thus, for example, the average octahedral $W-O$ bond is 0.15 Å longer than the average tetrahedral $W-O$ bond. This is not unexpected, considering the larger repulsion by the neighbouring electron pairs in the case of larger coordination.

Returning to the structural peculiarities of the Me_2XO_4 salts, Szpiridonov (1972) proposed the following structural formula:

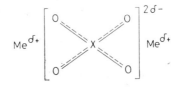

As in the classical structural formula, the central atom is hexavalent, having four bonds with a bond order of 1.5. The following merits of using this formula have been pointed out:

(*i*) it reflects the autonomous character of the XO_4 group;
(*ii*) it indicates that the XO_4 group remains unchanged in transition from one phase into another;
(*iii*) it shows the equality of the $X-O$ bonds;
(*iv*) it emphasizes the specific properties of the $Me-O$ linkages (e.g., the lengthening of the $Me-O$ bonds compared with those of hydroxides is consistent with the increase in the coordination number).

Similar reasoning has already been used in discussing the structures of mixed halides (p. 132).

It has already been mentioned that the XO_4 group makes the dominant contribution to creating the electron scattering pattern of the salts of oxyacids (Me_2XO_4). The two main reasons are the small scattering power of the alkali metal atoms of small atomic number and the large vibrational amplitudes associated with the $Me-O$ bonds. Accordingly, it is difficult, if not impossible, to determine certain geometrical features. It has also been argued that smaller amplitudes are to be expected for bonds with lower polarity (Tolmachev and Rambidi, 1972). Accordingly, compounds have been studied with heavier metal atoms and less polar $Me-O$ bonds in order to elucidate the molecular geometry more accurately.

The electron diffraction study of *thallium perrhenate*, $TlReO_4$, by Roddatis, Tolmachev, Ugarov, Ezhov and Rambidi (1974) yielded the following internuclear distances and mean amplitudes of vibration:

$r(Re-O) = 1.72 \pm 0.01$ Å $l(Re-O) = 0.04 \pm 0.02$ Å
$r(Tl-O) = 2.46 \pm 0.07$ Å $l(Tl-O) = 0.22 \pm 0.06$ Å
$r(Tl\ldots Re) = 3.21 \pm 0.02$ Å $l(Tl\ldots Re) = 0.18 \pm 0.02$ Å
$r(Tl\ldots O) = 4.44$ Å $l(Tl\ldots O) = 0.31 \pm 0.16$ Å.

The molecular configuration was described by a model in which a straight line connects the rhenium atom, the middle of one of the edges of the ReO_4 tetrahedron and the thallium atom:

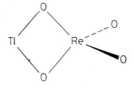

The average values for the bond angles are:

$$<O-Re-O = 98 \pm 6°$$
$$<O-Tl-O = 64 \pm 3°.$$

The molecular parameters of *indium molybdate*, In_2MoO_4 (Tolmachev and Rambidi, 1971), and *thallium molybdate*, Tl_2MoO_4 (Tolmachev and Rambidi, 1972), are given in Table 34. The main

TABLE 34

Molecular parameters (cf., Fig. 42)
for In₂MoO₄ and Tl₂MoO₄

		In_2MoO_4*	Tl_2MoO_4**
$r(Mo-O)$,	Å	1.81 ± 0.02	1.81 ± 0.02
$r(Me-O)$,	Å	2.20 ± 0.02	2.30 ± 0.03
$<O-Mo-O$		$85 \pm 5°$	$90 \pm 8°$
$<O-Me-O$		$68 \pm 3°$	$68 \pm 4°$
$l(Mo-O)$,	Å	0.10	0.10 ± 0.02
$l(Me-O)$,	Å	0.18	0.20 ± 0.03

* Tolmachev and Rambidi (1971).
** Tolmachev and Rambidi (1972).

parameters for thallium sulphate, Tl_2SO_4 (Ugarov, Ezhov and Rambidi, 1973b), are as follows:

$$r_g(S-O) = 1.48 \pm 0.02 \text{ Å} \qquad l_g(S-O) = 0.05 \pm 0.02 \text{ Å}$$
$$r_g(Tl-O) = 2.41 \pm 0.02 \text{ Å} \qquad l_g(Tl-O) = 0.14 \pm 0.02 \text{ Å}$$
$$<O-S-O = 107.5 \pm 4°$$
$$<O-Tl-O = 59 \pm 2.5°.$$

The overall molecular geometry shows complete analogy with those of K_2SO_4 and Cs_2SO_4, including a regular tetrahedral SO_4 group.

The configurations of the two molybdate derivatives are also well represented by Fig. 42. The MoO_4 tetrahedron, however, is somewhat distorted. It is interesting to compare the $Mo-O$ and $Me-O$ bonds of the molybdates with those in analogous oxides (cf., Table 35). The value of $r(Mo-O) = 1.81$ Å determined in the salts is intermediate between those of the terminal and bridging bonds of molybdenum trioxide. The $Me-O$ bonds ($Me = In$ or Tl) in the molybdates are 0.15–0.20 Å longer than those in corresponding oxides. The lengthening of the $Tl-O$ bond is even greater in Tl_2SO_4. Similarly, the mean amplitude of vibration associated with this bond is very large, in fact about twice as large as that in the analogous oxide molecule (Tolmachev and Rambidi, 1971, 1972). These results can be considered to indicate the validity of the structural formula shown on p. 152 for these compounds as well. The main characteristic common to these structures, which has been stressed, is the rigidity of the XO_4 group,

TABLE 35

Comparison of bond distances in molybdates and oxides

Compounds	Bonds	Distances, Å	References
$(MoO_3)_3$	Mo=O	1.67 ± 0.01	(a)
	Mo−O	1.89 ± 0.01	
In_2MoO_4	Mo−O	1.81 ± 0.02	(b)
	In−O	2.20 ± 0.02	
In_2O	In−O	2.00 ± 0.02	(c)
Tl_2MoO_4	Mo−O	1.81 ± 0.02	(d)
	Tl−O	2.30 ± 0.03	
Tl_2O	Tl−O	2.15 ± 0.02	(c)

(a) Egorova and Rambidi (1972).
(b) Tolmachev and Rambidi (1971).
(c) Tolmachev (1970).
(d) Tolmachev and Rambidi (1972).

whose geometry remains unchanged under transition from one phase to another. This geometry is also independent of the metal atom that is linked to it.

General as this description may be, a recent electron diffraction investigation on *barium tungstate*, $BaWO_4$, by Ivanov, Spiridonov, Erokhin and Levitskii (1973) showed some inconsistencies with this model. The WO_4 fragment was found to be considerably distorted from a regular tetrahedral configuration. The following internuclear distances were determined:

$$r(W-O) \quad = 1.82 \text{ Å}$$
$$r(Ba-O) \quad = 2.18 \text{ Å}$$
$$r(Ba \ldots W) = 3.18 \text{ Å}.$$

The bond angles O−W−O = 83° and O−Ba−O = 67° were also reported, based on the assumption of planarity for the ring

alculated from the above *r* values.

155

POLYMERIC OXIDES

The structures considered here are associates of two or more molecules. Except for the dinitrogen oxides and disulphur dioxide, the linkages between the molecular units are established by oxygen bridges. The emerging ring systems contain longer intra-ring bridging bonds and shorter terminal bonds. The accurate determination of the molecular geometry is hindered in some instances by the fact that there are several species present in the vapours or that the composition is not known. Accordingly, analysis of the vapour composition is an integral part of the structural determination. Electron diffraction, the technique used almost exclusively for elucidating the molecular geometries of polymeric oxides, requires at least a large relative abundance of the molecules to be studied. On the other hand, microwave spectroscopy may be applicable for species that occur as only a small fraction of the total vapour.

NITROGEN AND SULPHUR OXIDES

The most detailed studies have been performed on *dinitrogen tetroxide,* N_2O_4. The early sector-microphotometer electron diffraction study of Smith and Hedberg (1956) established a planar molecular configuration with D_{2h} symmetry (Fig. 43) and a particularly long $N-N$ bond. As an experimental scale error was later discovered in the reinvestigation of NO_2 (Blank, 1964), originally studied (Severinsson, 1955) at the same time as N_2O_4, a new determination of the structure of N_2O_4 was performed by McClelland, Gundersen and Hedberg (1972). The latter work verified the most important findings of the earlier study concerning the shape of the molecule. The size of the molecule was found to be 0.9% larger than previously reported. The parameters are shown in Tables 36 and 37.

The molecular configurations of dinitrogen tetroxide were found to be coplanar with D_{2h} symmetry both in the solid (Groth, 1963;

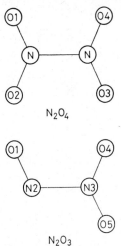

Fig. 43. The labelling of atoms for dinitrogen tetroxide and dinitrogen trioxide

TABLE 36

Internuclear distances and bond angles in nitrogen dioxide and dinitrogen tetroxide as determined by electron diffraction and in dinitrogen trioxide as determined by microwave spectroscopy

	NO_2 (a)	N_2O_4 (b)
$r(N-O)$, Å	1.202 ± 0.003	1.190 ± 0.002
$r(N-N)$, Å		1.782 ± 0.008
$<O-N-O$	$134.0 \pm 1.3°$	$135.4 \pm 0.6°$

		N_2O_3 (c)	
$r(N2-N3)$, Å	1.864	$<O1-N2-N3$	105.1°
$r(N2-O1)$, Å	1.142	$<O4-N3-N2$	112.7°
$r(N3-O4)$, Å	1.202	$<O5-N3-N2$	117.5°
$r(N3-O5)$, Å	1.217		

(a) Blank (1964).
(b) McClelland, Gundersen and Hedberg (1972).
(c) Brittain, Cox and Kuczkowski (1969).

Cartwright and Robertson, 1966) and liquid (Begun and Fletcher, 1960) phases. The main parameters reported by Groth (1963) from his X-ray diffraction study are $r(N-O) = 1.21$ Å, $r(N-N) = 1.75$ Å and $O-N-O$ bond angle $= 135°$.

The electron diffraction results on the molecular configuration of N_2O_4 are in good agreement with the vibrational spectra obtained from gaseous samples (Begun and Fletcher, 1960; Snyder and Hisatsune, 1957). However, the spectra obtained in the solid phase at the temperature of liquid nitrogen (Hisatsune, Devlin and Wada, 1960) and also those produced by the matrix isolation technique (Fately, Bent and Crawford, 1959), indicated the presence of both O_2N-NO_2 molecules with D_{2d} symmetry and O_2N-ONO configurations.

A molecular beam deflection study combined with mass spectrometric analysis by Novick, Howard and Klemperer (1972, 1974) established extensive polymerization beyond N_2O_4 in the vapour of the $NO_2-N_2O_4$ system. The higher polymers were observed to be non-polar and the following planar model with C_{3v} symmetry was suggested for the trimeric molecules:

$$
\begin{array}{c}
O \\
\parallel \\
N \\
O \diagup \quad \diagdown O \\
\\
O = N \qquad N = O \\
\diagup \quad \diagdown \quad \diagup \\
O \qquad O
\end{array}
$$

As Novick, Howard and Klemperer (1972) argued, the local symmetry of the above model is analogous to the structure of iso-dinitrogen tetroxide:

$$
\begin{array}{c}
O \qquad\qquad O \\
\diagdown \qquad\qquad \vdots \\
N \qquad\qquad N \\
\diagup \quad \diagdown \quad \diagup \quad \diagdown \\
O \qquad O \qquad\qquad O
\end{array}
$$

The conclusion originated from the fact that the dimeric nitrogen dioxide molecules detected under the experimental conditions ap-

TABLE 37

Mean amplitudes of vibration (l, Å) as determined by electron diffraction (ED) and as calculated from spectroscopic data (SP) for dinitrogen tetroxide

Atomic pairs*	ED	SP
N−O	0.0381 ± 0.0019	0.0391
N−N	0.0816 ± 0.0178	0.0543
N...O	0.0729 ± 0.0061	0.0670
O1...O2	0.0493 ± 0.0040	0.0525
O1...O3	0.0730 ± 0.0114	0.0711
O1...O4	0.0970 ± 0.0167	0.0943

* The numbering of the oxygen atoms is given in Fig. 43.

ED: the temperature of the electron diffraction experiment was −21°C, McClelland, Gundersen and Hedberg (1972).

SP: $T = -20$°C, Cyvin (1970).

peared to be polar. An alternative structure for the trimeric species was suggested by Liebman (1974):

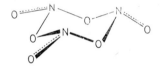

although it is expected to have a dipole moment.

The geometrical parameters of nitrogen dioxide are given in Table 36. Comparison of the data on NO_2 and $(NO_2)_2$ reveals no significant difference between the structures of the nitro groups in dinitrogen tetroxide and nitrogen dioxide. This is not surprising, however, in the light of the relatively weak linkage indicated by the long N−N bond between the two nitro groups of the N_2O_4 molecule. The N−N bond is 0.3 Å shorter in dinitrogen tetrafluoride, N_2F_4. Accordingly, it is more striking that the F−N−F bond angle in N_2F_4 (103.1 ± 0.6°; Gilbert, Gundersen and Hedberg, 1972) is almost the same as that in the NF_2 radical (102.5°; Bohn and Bauer, 1967) or in the NF_3 molecule (102.4°; Otake, Matsumura and Morino, 1968). For the sake of comparison, it is worthwhile to mention that both *anti* and *gauche* forms (see Fig. 44) appear in the vapour of dinitrogen tetrafluoride, as observed by Cardillo and Bauer (1969) and Gilbert, Gundersen and Hedberg (1972).

159

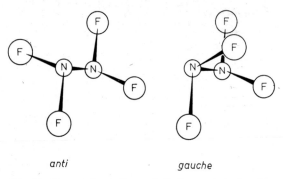

anti   gauche

Fig. 44. The conformations of dinitrogen tetrafluoride

The very long $N-N$ bond is undoubtedly the most interesting feature of the structure of dinitrogen tetroxide. In this connection, the relatively large amplitude of vibration associated with this bond is also noteworthy as it provides additional evidence for the looseness of the linkage between the two NO_2 groups. The calculated mean amplitudes of vibration from the spectroscopic analysis are in good agreement with the electron diffraction data, except for the value of $l(N-N)$ (see Table 37). Accordingly, some modification of the data on the molecular force field seems to be required. The mean amplitude of vibration for the essentially single $N-N$ bond (1.492 Å) in dinitrogen tetrafluoride is 0.048 ± 0.005 Å (Gilbert, Gundersen and Hedberg, 1972), which is thus much smaller than that determined in dinitrogen tetroxide.

The geometrical features of the N_2O_4 molecule inspired many workers to suggest various models and explanations. The original approach of Smith and Hedberg (1956) was to consider the $N-N$ bond to be of largely the π-type rather than the σ-type in order to account for the coplanarity of the molecule. The length and hence the weakness of the bond was thought to originate from the poor overlap properties of the orbitals in the bond direction. Further theoretical considerations are reviewed in the paper of McClelland, Gundersen and Hedberg (1972).

A complete determination of the molecular geometry of *dinitrogen trioxide*, N_2O_3, was carried out by microwave spectroscopy by Brittain, Cox and Kuczkowski (1969) using seven isotopic species. This work was preceded by an earlier attempt (Kuczkowski, 1965) that yielded partial structural information. The bond distances and bond angles for the planar configuration (Fig. 43) are presented in Table 36. It is interesting to observe again the coplanarity of the molecule (and a high barrier to internal rotation) despite the extraordinarily long central bond. In fact, the $N-N$ bond in N_2O_3 is even longer (by about 0.1 Å) than that in N_2O_4. A slight modification of the structure of the NO_2 group is observed compared with NO_2 itself. This group has a significant tilt of 2.4° towards the NO group, whereas a smaller tilt in the opposite direction was noted for nitric acid (cf., p. 139). A discussion of the bonding peculiarities was given by Brittain, Cox and Kuczkowski (1969).

The available data on the molecular geometry of *nitric oxide* dimer, $(NO)_2$, are rather uncertain although a series of both experimental and theoretical studies have been performed. Both solid-state X-ray diffraction (Lipscomb, Wang, May and Lippert, 1961) and gas-phase infrared spectroscopic (Dinerman and Ewing, 1970) studies established planar *cis* configurations. Assuming $r(N-O) = 1.15$ Å, Dinerman and Ewing (1970) determined $r(N-N) = 1.75$ Å and the $N-N-O$ bond angle = 90°. The X-ray diffraction data were as follows: $r(N-N) = 2.18 \pm 0.06$ Å, $r(N-O) = 1.12 \pm 0.02$ Å and $N-N-O$ bond angle = $101 \pm 3°$. The uncertainties of the spectroscopic results were estimated to be large enough to include the parameters obtained by X-ray diffraction. Various semi-empirical (Williams and Murrell, 1971; Vladimiroff, 1972) and *ab initio* (Skancke and Boggs, 1973) calculations were made and the results for the parameters reported covered a wide range. The recent work of Skaarup, Boggs and Skancke (1974) seems to be the most complete so far in which the structures and relative energies of different isomers have been determined by using the force method of Pulay (1969). These *ab initio* calculations may well facilitate further experimental work. The calculated internuclear distances and mean amplitudes of vibration for the two stable isomers found for $(NO)_2$ are given below:

	cis		*trans*	
	r, Å	l, Å	r, Å	l, Å
$N-N$	1.7674	0.0651	1.6861	0.0675
$N-O$	1.1623	0.0358	1.1704	0.0363
$N \ldots O$	2.3770	0.0607	2.3259	0.0762
$O \ldots O$	2.4326	0.0766	3.2737	0.0617

The dimeric form of *sulphur monoxide*, $(SO)_2$, has been observed only recently in a discharge system of sulphur dioxide using microwave spectroscopy (Tiemann, Lovas and Johnson, 1974; Lovas, Tiemann and Johnson, 1974). A planar *cis* configuration with the parameters

$$
\begin{aligned}
r(S-O) &= 1.458 \pm 0.002 \text{ Å} \\
r(S-S) &= 2.0245 \pm 0.0006 \text{ Å} \\
<S-S-O &= 112.7 \pm 0.5°
\end{aligned}
$$

was determined. Of these parameters the $S-S$ bond distance was obtained as an r_s parameter. Comparing the geometrical features of $(SO)_2$ with those of relevant systems, Tiemann, Lovas and Johnson (1974) noted, among others, the following geometrical variations. The sulphur–oxygen bond in the dimeric molecule is shorter than that in the monomer (1.48108 ± 0.00008 Å; Amano, Hirota and Morino, 1967), and is intermediate between those in S_2O (1.4637 ± 0.0005 Å; Tiemann, Hoeft, Lovas and Johnson, 1974) and SO_2 (*e.g.*, $r_e = 1.43076 \pm 0.00013$ Å; Saito, 1963). The sulphur–sulphur bond distance is similar to the $r(S-S)$ parameters found in HSSH (2.055 ± 0.001 Å; Winnewisser, 1972) and CH_3SSCH_3 (2.038 Å; Sutter, Dreizler and Rudolph, 1965) for what are considered to be typical single bonds.

A_4O_6 (A = P, PO, As or Sb) COMPOUNDS

The visual technique of electron diffraction was applied in the studies of the molecular geometry of the dimers of *diphosphorus trioxide*, P_4O_6, *diphosphorus pentoxide*, P_4O_{10}, and *diarsene trioxide*, As_4O_6 (Maxwell, Hendricks and Deming, 1937; Hampson and Stosick, 1938). For the arsene compound, another reinvestigation was also carried out (Lu and Donohue, 1944). Early sector-microphotometer studies have been reported on P_4O_{10} (Akishin, Rambidi and Zasorin, 1959) and *diantimony trioxide*, Sb_4O_6 (Akishin and Spiridonov, 1961b). The modern sector-microphotometer technique has been employed to elucidate the structures of P_4O_6 (Beagley, Cruickshank, Hewitt and Jost, 1969) and P_4O_{10} (Beagley, Cruickshank, Hewitt and Haaland, 1967).

All of the dimers listed above can be described by the general formula A_4O_6, where A may be P, As, Sb or the P=O group. The molecules have a cage-like structure with T_d symmetry in the vapour

162

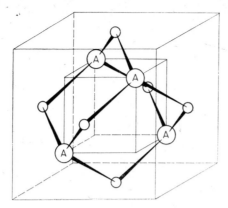

Fig. 45. The molecular configuration of A_4O_6
molecules (A = P, PO, As or Sb)

phase, according to both electron diffraction and vibrational spectroscopic evidence (for the latter see, *e.g.*, Beattie, Livingston, Ozin and Reynolds, 1970; Brumbach and Rosenblatt, 1972).

Because of their high symmetry, the geometry of these molecules can be described by a very small number of independent parameters, which is very advantageous for electron diffraction structural analysis. The configuration of the A_4O_6 molecule can be represented by means of two imaginary cubes, one inserted into the other. The atoms A are located at four of the corners of the inner cube in tetrahedral positions. The oxygen atoms are placed at the centres of the faces of the outer cube, as illustrated by Fig. 45. To describe the geometry of P_4O_{10}, a third cube is needed in order to accommodate the four terminal oxygen atoms at four of its corners in tetrahedral positions.

As expected, the A—O bond distance of the A_4O_6 molecules increases from A = P towards A = Sb. Parallel to this phenomenon, the A—O—A bond angles open slightly. The uncertainties of the parameters are relatively large:

$$P_4O_6 \quad r(P-O) = 1.638 \pm 0.003 \text{ Å}, \quad <P-O-P \quad = 126.4 \pm 0.7°$$
$$As_4O_6 \quad r(As-O) = 1.78 \pm 0.02 \text{ Å}, \quad <As-O-As = 128 \pm 2°$$
$$Sb_4O_6 \quad r(Sb-O) = 2.00 \pm 0.02 \text{ Å}, \quad <Sb-O-Sb = 129 \pm 2.5°.$$

The terminal P—O bond in P_4O_{10} (1.429 $\pm$ 0.004 Å) is considerably shorter than the bridging P—O bond (1.604 $\pm$ 0.003 Å). The terminal

11*

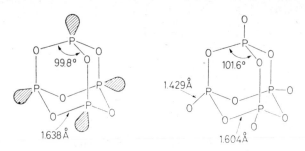

Fig. 46. The molecular models and geometrical parameters of P_4O_6 (Beagley, Cruickshank, Hewitt and Jost, 1969) and P_4O_{10} (Beagley, Cruickshank, Hewitt and Haaland, 1967)

bond is barely longer than that which could be calculated for the double bond, whereas the bridging bond seems to be an ordinary single bond (cf., Cruickshank, 1961).

Probably the most interesting structural variation to be observed in comparing the geometries of P_4O_6 and P_4O_{10} is the change as the lone pair of electrons on the phosphorus atom is substituted by the $P=O$ bond. As the data in Fig. 46 show, the bridging $P-O$ bond shortens significantly and the intra-ring $O-P-O$ angle opens slightly. Although the V.S.E.P.R. model does not decide in general whether a lone pair or a double bond exercises more repulsion towards the neighbouring electron pairs, the structural data on simple phosphorus compounds indicate that the stronger repulsion can be attributed to the lone pair of electrons. A similar observation has been made for the sulphur derivatives (Hargittai and Mijlhoff, 1973). To illustrate this point, structural data for compound pairs of PX_3 and POX_3 (X = F, Cl or Br) are presented in Table 38, showing consistency with the variations observed for P_4O_6 and P_4O_{10}.

The bond distances and bond angles in phosphorus oxides with general formula P_4O_{6+n}, where n = 0, 2, 3 or 4, were compared by Beagley, Cruickshank, Hewitt and Jost (1969). The molecular structures of P_4O_8 (Beagley, Cruickshank, Hewitt and Jost, 1969) and P_4O_9 (Beagley, Cruickshank, Hewitt and Haaland, 1967) have been determined in the crystal phase by X-ray diffraction. The geometrical parameters, and most notably the bond distances, show little variation in this compound series. The structural data of P_4O_7 were estimated from the weighted mean values found for other compounds. The

TABLE 38

Bond distances and bond angles in PX_3 and
POX_3 molecules (X = F, Cl or Br)

$r(P-X)$

F	1.570	Å	(a1)	1.524_0	Å	(a4)
Cl	2.040	Å	(a2)	1.993_1	Å	(a4)
Br	2.220_4	Å	(a3)	2.14	Å	(b)

$<X-P-X$

F	97.8°	(a1)	101.3°	(a4)
Cl	100.3°	(a2)	103.3°	(a4)
Br	101.0°	(a3)	105.6°	(b)

(a) Vapour-phase electron diffraction ; (1, 3, 4) r_g, $<_\alpha$; (2) r_a.
   (1) Morino, Kuchitsu and Moritani (1969).
   (2) Hedberg and Iwasaki (1960).
   (3) Kuchitsu, Shibata, Yokozeki and Matsumura (1971).
   (4) Moritani, Kuchitsu and Morino (1971). Results of a recent microwave spectroscopic study on $POCl_3$ by Li, Chen and Durig (1972): $r(P-Cl) = 1.989 \pm \pm 0.002$ Å and $<Cl-P-Cl = 103.7 \pm 0.2°$.
   (b) Crystal-phase X-ray diffraction results of Olie and Mijlhoff (1969).

weighting took into consideration the fact that the electron diffraction results were three to four times more accurate than the X-ray diffraction parameters.

Concerning the intramolecular motion of the molecules A_4O_6, it is relevant to recall that Hargittai and Hedberg (1972) pointed out the stiffness of the similarly cage-like, highly symmetrical structure of adamantane, $C_{10}H_{16}$ [or $(CH)_4(CH_2)_6$]. The stiffness is indicated by the relatively small values of the mean amplitudes of vibration associated with non-bonded atomic pairs C . . . C. The measured values were in good agreement with the parameters calculated from spectroscopic data (Cyvin, Cyvin and Brunvoll, 1972). The electron diffraction mean amplitudes of vibration reported for P_4O_6 and P_4O_8 in the case of some important non-bonded atomic pairs are considerably larger than the calculated values (Cyvin and Cyvin, 1971).

165

There are various species in the vapours of the trioxides of chromium, molybdenum and tungsten corresponding to various degrees of polymerization. It is necessary to know the composition of the vapour in order to determine the molecular geometry and the lack of such information may seriously hinder the structural analysis. According to the mass spectra of chromium trioxide (McDonald and Margrave, 1968), there are considerable amounts of several species, including trimers and tetramers, in the vapour phase under the experimental conditions used. The trimeric species prevail in the vapour of molybdenum trioxide with some tetrameric, pentameric and monomeric forms also present (Berkowitz, Inghram and Chupka, 1957). The relative abundances of the $(WO_3)_3$, $(WO_3)_4$ and $(WO_3)_5$ molecules appearing in the vapour of tungsten trioxide was found to be 568 : 168 : 1 (Berkowitz, Chupka and Inghram, 1957). These data, obtained at 1492 K, are in good agreement with the results of an earlier study (Ueno, 1941) corresponding to 84.9 and 15.1 mole % of trimer and tetramer, respectively, at 1393 K. Two more recent mass spectrometric studies (Ackerman and Rauh, 1963; Kazenas and Tsvetkov, 1967) provided expressions for the partial pressures of the gaseous sublimation products of solid $WO_{2.96}$:

$$\log p_i \, (\text{atm}) = \frac{a}{T} + b$$

with the numerical values of a and b presented in Table 39. The data referring to the vapour composition of tungsten trioxide have been presented above in somewhat more detail. Only the electron diffraction analysis of tungsten trioxide was performed with careful examination of the influence of the changes in the relative abundances of various species (Hargittai, 1970).

The electron diffraction patterns of the vapour of *chromium trioxide* were recorded at about 300°C. The mean length of the chromium–oxygen bonds was found to be 1.68 Å (Zasorin, Ivanov, Spiridonov, Hargittai and Hargittai, 1973). This value is consistent with the observation of Stephens and Cruickshank (1970a) that the Cr−O bond in chromium oxides and chromates lengthens as the degree of association increases. The data presented below may not show the variations from compound to compound to be significant, but nevertheless the trend indicated above seems to be present:

	Mean bond distance $r(Cr-O)$, Å	
Vapour phase K_2CrO_4	1.66	(Spiridonov and Lutoshkin, 1970)
Crystal phase $K_2Cr_2O_7$	1.67	(Brandon and Brown, 1968)
Crystal phase $(NH_4)_2Cr_2O_7$	1.67	(Stephens and Cruickshank, 1970b)
Vapour phase $(CrO_3)_3$ and $(CrO_3)_4$	1.68	(Zasorin, Ivanov, Spiridonov, Hargittai and Hargittai, 1973)
Crystal phase $(CrO_3)_\infty$	1.69	(Stephens and Cruickshank, 1970a)

Assuming that all of the polymeric species of chromium oxides in the vapour have ring structures, the values 1.78 and 1.58 Å were obtained for the bridging and terminal bonds. The experimental

TABLE 39

Coefficients in the expression log p_i (atm) $= \dfrac{a}{T} + b$ for the partial pressures of the vapour of solid $WO_{2.96}$

	Kazenas and Tsvetkov (1967)		Ackerman and Rauh (1963)	
	$T = 1300-1600$ K		$T = 1300-1500$ K	
	a	b	a	b
W_3O_9	−23800	11.76	−23934	14.891
W_3O_8	−24840*	11.76	−24830	14.969
W_4O_{12}	−27190	13.27	−27519	16.780
W_2O_6	−25180	11.81	−25521	14.194

*The value of − 14840 in the original paper by Kazenas and Tsvetkov (1967) is presumably a misprint

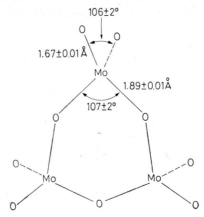

Fig. 47. The bond distances and bond angles
determined for a planar ring model of trimeric
molybdenum trioxide
(Egorova and Rambidi, 1972)

distributions could be well approximated by the theoretical curves calculated for either the trimeric or the tetrameric ring configurations, as well as their mixtures. More detailed structural information, however, can hardly be expected to be obtained by using the experimental data currently available because of the complexity of the vapour composition.

The temperature of the electron diffraction experiment on molybdenum trioxide was 1000°C (Egorova and Rambidi, 1972). The electron diffraction data were interpreted under the assumption that the vapour consisted of a ring trimeric species. The geometrical parameters obtained for a planar model with D_{3h} symmetry are shown in Fig. 47. The bond distances determined are consistent with the data for other molybdenum compounds (cf., Cotton and Wing, 1965).

Various models were suggested for the structures of the tungsten oxide molecules prior to the electron diffraction study (see, *e.g.*, Millner, 1955; Speiser and Pierre, 1964; Stull, 1966). The discovery of volatile complexes between tungsten oxides and water by Millner and Neugebauer (1949) stimulated interest in the structural chemistry of tungsten oxides. The matrix isolation spectra of the tungsten oxides were interpreted by means of non-planar ring models with C_{3v} symmetry for $(WO_3)_3$ and C_{4v} symmetry for $(WO_3)_4$ (Weltner and

McLeod, 1965). The assignment of the spectra, however, is rather uncertain in several instances because of the complexity of the composition of the vapour. The configurations of

and

were considered for the W_2O_6 and W_3O_8 molecules, respectively, which are also present in the vapour.

The electron diffraction data on the vapour of tungsten trioxide collected at a nozzle temperature of about 1400°C (Hargittai, Hargittai, Spiridonov and Erokhin, 1971) showed the trimeric species to predominate, in agreement with the results of the mass spectrometric analysis. The presence of other polymeric species could not be ruled out; on the contrary, the addition of a small amount of tetramers to the model considerably improved the agreement between the experimental and theoretical distributions (Hargittai, 1970). The results of the study of Hargittai, Hargittai, Spiridonov and Erokhin (1971) on the molecular geometry of the trimeric tungsten trioxide molecule can be summarized as follows. The $(WO_3)_3$ molecules have a six-membered ring structure with alternating tungsten and oxygen atoms. The mean tungsten–oxygen bond distance is 1.77 ± 0.02 Å. The experimental data could be approximated much better by a non-planar static model than by a planar ring skeleton. The deviation from planarity can be characterized by the puckering angle, ε, shown in Fig. 48. The individual values of the bridging and terminal W—O bond distances could not be determined, but it was possible to ascertain that the bridging bond is longer than the terminal bond. A series of models could be used to reproduce the experimental distributions

169

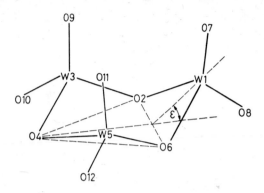

Fig. 48. The molecular model of trimeric tungsten trioxide with C_{3v} symmetry (ε is the puckering angle) and the numbering of atoms

containing various combinations of the two tungsten–oxygen bond distances. The geometrical parameters for two such models with different combinations of the bond distances are presented in Table 40.

A similar structure to that of $(WO_3)_3$ was established for crystalline $(SO_3)_3$ (Westrick and MacGillavry, 1941; Pascard and Pascard-Billy, 1965; McDonald and Cruickshank, 1967). A puckering angle of 29° (see ε, Fig. 48) can be calculated from the parameters for S_3O_9 given by McDonald and Cruickshank (1967). The puckering angle of 15–20°

TABLE 40

Bond angles and puckering angles in the models with C_{3v} symmetry of W_3O_9 as determined assuming two sets of values for the bond distances according to Hargittai, Hargittai, Spiridonov and Erokhin (1971)

Parameters*		Model I	Model II
$r(W1-O2)$	(assumed)	$\{$ 1.87 Å	1.82 Å $\}$
$r(W1-O7)$		1.68 Å	1.73 Å
$<O2-W1-O6$		95°	85°
$<O7-W1-O8$		109.5°	120°
$<W1-O2-W3$		135°	144°
Puckering angle (ε)		18°	16°

* The numbering of atoms is given in Fig. 48.

170

obtained for the W_3O_9 molecule seems to be relatively small, considering the high temperature of the vapour at which the electron diffraction pictures were taken. Hargittai, Hargittai, Spiridonov and Erokhin (1971) emphasized that the model with C_{3v} symmetry may be only one of a number of configurations of the six-membered ring structure of W_3O_9. It was also supposed that the structure might be described by a planar ring and D_{3h} symmetry with large shrinkage effects caused by the enormous bending vibrations expected at this temperature. It was also pointed out that, in the complicated thermal motion, probably a series of configurations occur. In order to examine whether an average structure free from shrinkage could be expected to be planar or nearly planar, the quantities $K = [\langle(\Delta x)^2\rangle + \langle(\Delta y)^2\rangle]/2r$ were calculated (Cyvin, Hargittai, Cyvin and Hargittai, 1975). Although the very approximate nature of these calculations has to be kept in mind, the results do not indicate that the shrinkage effect is the main origin of the apparent ring puckering of the W_3O_9 molecules, as was determined in the electron diffraction study.

TETRAMERIC SELENIUM TRIOXIDE

Mijlhoff (1964, 1965a, 1965b) has extensively studied the molecular structure of *selenium trioxide* in both the crystalline and vapour phases. According to the X-ray diffraction investigation, the molecules form tetrameric species in the crystal. Vapour pressure measurements indicated that the tetrameric form also predominated in the vapour,

TABLE 41

Bond lengths and bond angles of Se_4O_{12} in crystalline and gaseous phases

Parameters	Crystal Mijlhoff (1965a)	Gas Mijlhoff (1965b)
$r(Se-O)_b$	1.55 ± 0.012 Å	1.56 ± 0.02 Å
$r(Se-O)_t$	1.78 ± 0.013 Å	1.80 ± 0.02 Å
$< Se-O_b-Se$	$123.2 \pm 0.75°$	$121 \pm 2°$
$< O_b-Se-O_b$	$98.7 \pm 0.85°$	$95-104°$
$< O_t-Se-O_t$	$128.2 \pm 0.85°$	$115-128°$

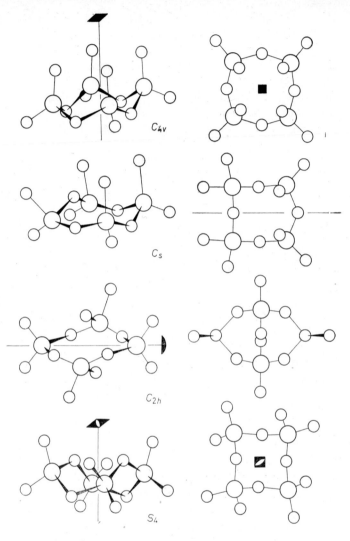

Fig. 49. Four ring conformations examined for the tetrameric selenium trioxide molecule (Mijlhoff, 1964)

although some of the polymeric molecules dissociated into monomers. The electron diffraction data were observed to be consistent with these findings. The temperature of the experiment was 120°C. The molecular models considered in detail for the tetrameric species are shown in Fig. 49. Of these, the best agreement with the experimental data was achieved for the model with S_4 symmetry. The structural parameters obtained in both the crystal- and vapour-phase studies are presented in Table 41. A more complete structural determination for the vapour-phase molecular geometry will probably require simultaneous mass spectrometric analysis.

$$[(CH_3)_2AlOCH_3]_3$$

The results of the structural elucidation of *dimethylaluminium methoxide* trimer, $[(CH_3)_2AlOCH_3]_3$, are also considered here as the linkage between the monomeric units is established through bridging oxygen atoms similar to those found in the polymeric oxides. The results of the electron diffraction study by Drew, Haaland and Weidlein (1973) are given in Table 42 and the molecular model is shown in Fig. 50. The parameters presented were obtained by assuming a non-planar ring configuration with C_{3v} symmetry. Less satisfactory agreement with the experimental data was achieved for a structural refinement

TABLE 42

Bond lengths (r_a parameters) and bond angles of dimethylaluminium methoxide trimer, $[(CH_3)_2AlOCH_3]_3$, determined from the electron diffraction data for a model with C_{3v} symmetry by Drew, Haaland and Weidlein (1973)

Bond lengths, Å		Bond angles	
$r(C-H)$	1.106 ± 0.002	$<Al-C-H$	$113.5 \pm 1.7°$
$r(Al-C)$	1.957 ± 0.003	$<O-C-H$	$105.6 \pm 1.6°$
$r(Al-O)$	1.851 ± 0.003	$<C-Al-C$	$117.3 \pm 0.8°$
$r(O-C)$	1.436 ± 0.003	$<O-Al-O$	$103.2 \pm 1.1°$
		$<Al-O-Al$	$125.8 \pm 0.4°$

173

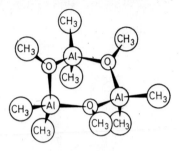

Fig. 50. Dimethylaluminium methoxide
trimer, [(CH₃)₂AlOCH₃]₃

based on a planar ring model. The Al_3O_3 ring skeleton was also shown to be non-planar by spectroscopic studies (Mann, Haaland and Weidlein, 1973). One of the interesting features of the molecular geometry is the coplanarity of the three bonds of the oxygen atom.

HYDROGEN-BONDED COMPLEXES

Extensive literature is available on the systems with hydrogen bonds (see, *e.g.*, Vinogradov and Linnel, 1971), but references are scarce on vapour-phase studies of the geometries of hydrogen-bonded molecules. The hydrogen bond may be either intra- or intermolecular, and this discussion is concerned with hydrogen-bonded molecules in which "monomeric" molecules are connected with the hydrogen bond.

DIMERS OF ORGANIC ACIDS

The dimeric molecules of organic acids are formed from the monomers through hydrogen bonds. The vapour-phase structural determinations are aimed firstly at measuring the length of the $O-H...O$ hydrogen bond and at establishing the changes in the geometry as compared with the monomers. So far, the molecular geometries of dimeric and monomeric *formic acid, acetic acid* and *propionic acid* have been elucidated in the vapour phase. The most recent investigations of formic acid and acetic acid, together with the study on propionic acid, were based on experimental data collected in the Oslo electron diffraction apparatus (formic acid, Almenningen, Bastiansen and Motzfeldt, 1969; acetic acid, Derissen, 1971a; propionic acid, Derissen, 1971b).

Previous studies on the molecular structure of monomer and dimer formic acid were reviewed by Almenningen, Bastiansen and Motzfeldt (1969). In addition to the references cited, Bonham and Su (1968) reported the results of another electron diffraction study with parameters different from those of Almenningen, Bastiansen and Motzfeldt (1969). The ring structure and the formation of two hydrogen bonds had already been established by the early work of Pauling and Brockway (1934). The arrangement is as follows:

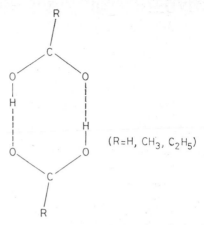

(R=H, CH_3, C_2H_5)

The discovery of the two types of carbon–oxygen bonds in the ring, corresponding to double and single bonds, was a later result. The geometrical parameters of both dimeric and monomeric formic acid (Almenningen, Bastiansen and Motzfeldt, 1969) are given in Table 43. The data on acetic acid and propionic acid from the papers of Derissen (1971a, 1971b) are also presented.

The dimeric formic acid molecule appears to have a strictly coplanar configuration, differing from earlier models. The presence of non-ring structures in the vapour cannot be excluded, however.

The single C−O bond in the dimer is considerably shorter than that in the monomer in all three systems (cf., Table 43). On the other hand, the C=O bond is much less influenced by the formation of the hydrogen bond. This is consistent with the double bond having a relatively large force constant (Almenningen, Bastiansen and Motzfeldt, 1969).

Ab initio calculations have recently been performed (Flood, 1974) on monomeric and dimeric formic acid in order to examine the correlation between the variations of the lengths of the single and double carbon–oxygen bonds and the changes in the electronic structure when the carboxylic group dimerizes.

The O−C−O bond angle is larger in dimeric than in monomeric formic acid. The differences between the analogous bond angles in the other two systems are within experimental error. The O−H bond is longer in dimeric than in monomeric formic acid. This bond is again strongly influenced by the formation of the hydrogen bond. The values of $r(O−H)$ determined for formic acid were used as assumed

TABLE 43

Geometrical parameters for monomeric and dimeric carboxylic acids as determined by electron diffraction

	Formic acid (a)		Acetic acid (b)		Propionic acid (c)	
	Monomer HCOOH	Dimer $(HCOOH)_2$	Monomer CH_3COOH	Dimer $(CH_3COOH)_2$	Monomer CH_3CH_2COOH	Dimer $(CH_3CH_2COOH)_2$
$r(C{=}O)$, Å	1.217±0.0026	1.220±0.0029	1.214±0.003	1.231±0.003	1.211±0.003	1.232±0.006
$r(C{-}O)$, Å	1.361±0.0030	1.323±0.0029	1.364±0.003	1.334±0.004	1.367±0.004	1.329±0.008
$r(O{-}H)$, Å	0.984±0.024	1.036±0.017	0.97*	1.03*	0.97*	1.03*
$r(O{-}H{\ldots}O)$, Å		2.703±0.0070		2.680±0.008		2.711±0.014
$<O{-}C{-}O$	123.4±0.46°	126.2±0.48°	122.8±0.6°	123.4±0.8°	122.1±0.8°	123.7±1.6°
Temperature	175°C	12°C	160°C	24°C	215°C	55°C

* Assumed parameters.
(a) Almenningen, Bastiansen and Motzfeldt (1969).
(b) Derissen (1971a).
(c) Derissen (1971b).

parameters in the structural analyses of acetic acid and propionic acid. The variations of the bond distances of formic acid are consistent when the dimer and monomer are compared. At least, this is true when considering the trends in the changes compared with the differences in the respective stretching force constants obtained from spectroscopic data (Nakamoto and Kishida, 1964; Kishida and Nakamoto, 1964):

$$k(\times 10^5 \text{ dyne cm}^{-1})$$

	HCOOH	(HCOOH)$_2$
C=O	11.20	10.00
C−O	4.60	5.50
O−H	6.90	4.70
O...H	−	0.36

The small value for the force constant of the O . . . H linkage seems to correlate well with the relatively large mean amplitude of vibration associated with it, as obtained from electron diffraction measurements (0.126 ± 0.0045 Å; Almenningen, Bastiansen and Motzfeldt, 1969) and calculated from spectroscopic data (0.115 Å; Alfheim, Hagen and Cyvin, 1971).

The influence of isotopic substitution on the hydrogen bond in dimeric formic acid has also been examined by Almenningen, Bastiansen and Motzfeldt (1970). The hydrogen bond was found to be 0.02 Å longer in (HCOOD)$_2$ than in (HCOOH)$_2$, which is in good agreement with the findings of crystal-phase studies on isotopic effects on the hydrogen bond (Delaplane and Ibers, 1969).

There are important differences between the structures of dimeric organic acids in the vapour and crystal phases. The most striking is that instead of the ring configurations predominating in the vapours, in the crystals the molecules are linked to each other to form a chain. Further, the hydrogen bonds observed in the crystal are shorter for all three systems. These are presented below for comparison with the vapour-phase data given in Table 43:

	$r(O−H . . . O)$	
Formic acid	2.58 ± 0.03 Å	(Holtzberg, Post and Fankuchen, 1953)
Acetic acid	2.626± 0.006 Å	(Nahringbauer, 1970)
Propionic acid	2.644 Å	(Strieter, Templeton, Scheuerman and Sass, 1962)

BIMOLECULES

The dimeric organic acid structures discussed above could not be studied by microwave spectroscopy as these symmetric configurations have no permanent electric dipole moment. On the other hand, when two different molecules are linked together by a hydrogen bond, forming the so-called bimolecules, they possess a dipole moment and, accordingly, their structures can be studied by microwave spectroscopy. As the bimolecules are obtained, there are usually monomers and dimers also present in the vapour, making the spectroscopic study more difficult. (For such a complicated system, an electron diffraction study would be hopeless at the outset.)

Trifluoroacetic acid–formic acid, $F_3CCOOH - HCOOH$:

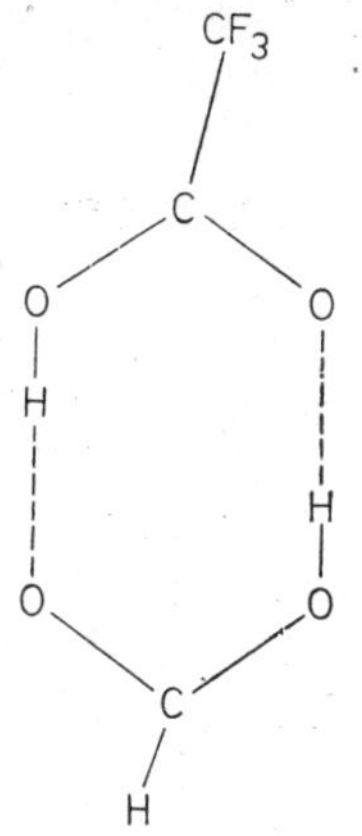

was the first bimolecule whose molecular geometry was studied by microwave spectroscopy (Costain and Srivastava, 1961). Later, structural determinations on trifluoroacetic acid–acetic acid, $F_3CCOOH - CH_3COOH$, and trifluoroacetic acid–monofluoroacetic acid, $F_3CCOOH - FCH_2COOH$, were reported (Costain and Srivastava, 1964). The primary interest centred around elucidating the length of the hydrogen bond in all three systems. Most of the geometrical parameters, especially those for the monomers, were assumed from other studies. The influence of isotopic substitution on the hydrogen bond has also been studied:

12*

179

	$r(O-H...O)$, Å	$r(O-D...O) - r(O-H...O)$, Å
$F_3CCOOH - HCOOH$	2.69	0.011
$F_3CCOOH - CH_3COOH$	2.67	0.012
$F_3CCOOH - FCH_2COOH$	2.69	0.010

As the above parameters refer to the ground vibrational state, it seems to be reasonable to consider the $r(O-D...O) - r(O-H...O)$ difference as a consequence of the difference in the zero-point vibrations of the deuterated and non-deuterated species (Costain and Srivastava, 1964). Otherwise, the hydrogen bonds in the bimolecules are observed to have very nearly the same lengths as those in the dimeric acid molecules. As several microwave spectroscopic studies (Srivastava and Goyal, 1968a, 1968b, 1972; Gantam, 1970) have dealt with bimolecules, although without yielding geometrical data, it is expected that more detailed structural determinations will eventually appear.

POLYMERIC HYDROGEN FLUORIDE

The composition of the vapour of *hydrogen fluoride* is complicated. The molecular beam deflection and mass spectrometric studies of Dyke, Howard and Klemperer (1972) showed that the polymerization of hydrogen fluoride is continuous, *i.e.*, there are many more species in the vapour in addition to monomers, dimers and hexamers than had been suggested, and that the hydrogen fluorides with a higher degree of polymerization are non-polar. The latter observation strongly suggests a ring configuration, as the non-rigidity of the chain structures is not thought to be the cause of the absence of a dipole moment.

The bond is 0.917 Å long (r_e value) in monomeric hydrogen fluoride (Spanbauer, Rao and Jones, 1965; Webb and Rao, 1968). The structure of non-rigid dimeric molecules has been studied in detail by radio-frequency and microwave spectroscopy (Dyke, Howard and Klemperer, 1972). These results are consistent with the structure of the monomer and the following non-linear model:

The value of 2.79 Å, determined for the length of the F—H...F hydrogen bond agrees well with the result of *ab initio* calculations (2.78 Å) for $(HF)_2$ (Kollman and Allen, 1970).

In addition to the earlier visual electron diffraction study by Bauer, Beach and Simons (1939), Janzen and Bartell (1969) reported an up-to-date reinvestigation of vapour-phase hydrogen fluoride. The electron scattering patterns were recorded at two different temperatures ($-19°C$ and $22°C$) in the latter work in order to examine the influence of temperature on the degree of association. Also, very high vapour pressures were employed in order to facilitate polymerization. The electron diffraction data were interpreted in terms of monomeric and ring structure hexameric species. It was not possible to distinguish between non-rigid planar ring and real non-planar ring models. Because of the large-amplitude motion, even the "chair" and "boat" forms

"chair"   "boat"

of the non-planar model could not be distinguished. Janzen and Bartell (1969) argued that the $(HF)_6$ molecules alternate between the two forms in their thermal motion. The average value of the F—F—F angles was found to be 104°. The lengths of the F—H...F hydrogen bonds were reported to be 2.525 ± 0.003 Å and 2.533 ± 0.003 Å for the experimental temperatures of $-19°C$ and $22°C$, respectively.

A shift towards the formation of chain structures was observed by Reichert and Hartmann (1972) as the vapour pressure of hydrogen fluoride was reduced. Planar chain hexameric and heptameric species were found to predominate under the experimental conditions used (temperature $-70°C$, pressure 0.01 torr) in the microwave spectroscopic study. The molecular models proposed by Reichert and Hartmann (1972) can be characterized by the following fragments:

181

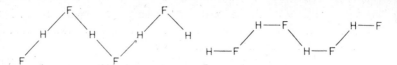

They also observed a shortening of the H−F bond and a lengthening of the hydrogen bond as a consequence of the formation of a heptamer in the reaction $(HF)_6 + HF$:

Hexamer: $r(H-F) = 0.9997$ Å and $r(H \ldots F) = 1.4998$ Å
Heptamer: $r(H-F) = 0.9640$ Å and $r(H \ldots F) = 1.6105$ Å.

Accordingly, the stability of the chain was shown to decrease in the transition from hexamer to heptamer. On the other hand, on the basis of the following data:

vapour-phase dimer: $r(F-H \ldots F) = 2.79$ Å (Dyke, Howard and Klemperer, 1972)
vapour-phase ring hexamer: $r(F-H \ldots F) = 2.53$ Å (Janzen and Bartell, 1969)
crystal-phase chain: $r(F-H \ldots F) = 2.49$ Å (Atoji and Lipscomb, 1954)

Dyke, Howard and Klemperer (1972) concluded that the hydrogen bond gradually shortens with increasing degree of polymerization.

In spite of the numerous studies, many important details of the structures of hydrogen fluoride are not yet known and further studies are needed in order to throw more light on the composition of the vapour and the geometries of the species present. The geometrical studies have been supplemented by spectroscopic investigations (Kittelberger and Hornig, 1967; Tubino and Zerbi, 1969; Cyvin, 1973b) that will certainly also be valuable for more detailed structural determinations. As the knowledge of the vapour composition is a prerequisite for more complete elucidation of the molecular geometry, interestingly enough, *ab initio* calculations may be instrumental in this respect. According to the results of Del Bene and Pople (1971): (1) for the polymers $(HF)_n$, where $n = 3, 4, 5$ or 6, the ring configurations are the most stable; (2) for dimers, the chain structure is more advantageous; (3) the high stability of the ring systems, particularly of the tetrameric, pentameric and hexameric ring molecules, indicates that a considerable proportion of these species is to be expected in the vapour phase in addition to the monomers and dimers.

Complete dissociation of *ammonium chloride*, NH_4Cl, vapour into ammonia and hydrogen chloride was observed in a mass spectrometric study (Goldfinger and Verhaegen, 1969). The electron diffraction data of Shibata (1970), however, indicated 60% dissociation only, leaving the remainder of the vapour as monomeric NH_4Cl species. Models with one, two or three hydrogen bonds have been tested against the measurements and are shown in Fig. 51. The configuration with one hydrogen bond was preferred. This result is supported by a recent matrix isolation infrared spectroscopic study by Ault and Pimentel (1973), in agreement with earlier *ab initio* calculations (Clementi, 1967a, 1967b). A similar conclusion was drawn from an electron diffraction investigation for ammonium fluoride, NH_4F (Shibata, 1970).

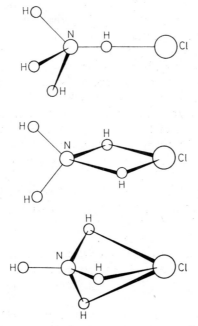

Fig. 51. Molecular models of ammonium
chloride, NH_4Cl

TRANSITION METAL COMPLEXES*

The molecular geometries of relatively few coordination compounds of transition metals have been determined in the vapour phase. The number of such investigations, however, has been rapidly increasing recently. It is also interesting to note that reinvestigations have often been initiated in order to gain a more detailed insight into the molecular structure as the experimental techniques improve or to decide controversial issues. Electron diffraction rather than microwave spectroscopy is the primary source of information on molecular geometry. This is partially explained by the high symmetry of most of the simpler transition metal complexes. The other reason is just the relatively large size of these molecules for a spectroscopic determination of their geometry. For the same reason, the electron diffraction studies are intended to answer specific questions only, rather than giving a complete structural determination. Different assumptions are used extensively in the structural analysis, for example, concerning the local symmetries of the atomic groups. These assumptions will be detailed in subsequent discussion only in specific instances in which they are felt to influence considerably the determined parameters or their uncertainties.

The bonding of transition metal complexes has been discussed in detail in numerous monographs and texts. Only the excellent book by Cotton and Wilkinson (1972) is mentioned here, and their classification is followed to some extent.

The *complexes of the π-acceptor ligands* include carbonyls, their substituted derivatives and PF_3 complexes. It is well known that there

---

* The compounds of non-transition and transition elements are not strictly separated in this work because of other considerations in the classification. Thus, transition metal compounds have already been discussed in other sections, *e.g.*, among the polymeric oxides. Numerous compounds of non-transition elements, on the other hand, will occur in the section on cyclopentadienyl complexes.

is a σ-bond formed in these complexes utilizing the lone pair of electrons on the ligand as donor and the vacant metal orbitals. This is supplemented by the π-bonding formed by the vacant orbitals of the ligands accepting electron density from non-bonding metal orbitals. This process is also called back-coordination and its extent may be very different in various compounds.

The *π-complexes* form the other large group of compounds dealt with in this chapter. Again, the ligands both donate electrons to and accept electrons from the metal atom. However, only the π-orbitals of the ligand are used and therefore there is only π-bonding in these complexes. The so-called sandwich compounds belong to this group, the most important representatives being the metallocenes containing cyclopentadienyl rings.

There are, of course, numerous intermediates between the two main classes briefly characterized above. The olefin complexes of transition metal carbonyls are among those discussed. The olefin−metal bonds in these compounds are of neither purely σ- nor purely π-character. The allyl complexes, represented here by π-allyl cobalt tricarbonyl, are intermediates themselves between the olefin complexes and the sandwich compounds. The bonding in the allyl complexes is characterized by the interaction between the metal atom and the π-electron density delocalized on the three carbon atoms.

In subsequent sections, attention is focused on the geometrical data. The characteristics of the bonding are considered only when direct correlation can be seen between them and the results of the vapour-phase molecular geometries.

COMPLEXES OF π-ACCEPTOR LIGANDS

MONONUCLEAR CARBONYLS

In early work, the visual technique of electron diffraction was applied by Brockway, Ewens and Lister (1938) to study the molecular structures of *chromium*, *molybdenum* and *tungsten hexacarbonyl*, $M(CO)_6$. Table 44 contains some parameters for $Cr(CO)_6$ from this study, together with more recent results for the other two compounds. The molecules were found to be of regular octahedral configuration (O_h symmetry), as shown by Fig. 52. In the course of a modern elec-

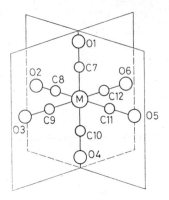

Fig. 52. The octahedral molecular geometry of M(CO)$_6$ carbonyls and the numbering of the atoms

tron diffraction structural analysis, Arnesen and Seip (1966) carried out a detailed study on the applicability of electron scattering functions in various approximations. The failure of the first Born approximation was shown for structural studies on molecules that consist of atoms with large differences in atomic numbers. The application of the first Born approximation may lead to the establishment of distorted structures, even for regular polyhedral molecules. The geometrical parameters of Mo(CO)$_6$ and W(CO)$_6$ were determined by Arnesen and Seip (1966) by utilizing complex electron scattering amplitudes obtained by the partial wave method (Seip, 1967). The comprehensive paper by Seip (1967) contains the final results, which differ little from those given in the earlier paper by Arnesen and Seip (1966). The parameters referring to bonds are given in Table 44, and those for non-bond distances are presented in Table 45, including the l values and shrinkages. The vibrational parameters were calculated for all three compounds from spectroscopic data by Brunvoll (1965). The expressions for both the linear and non-linear shrinkages* in the M(CO)$_6$-type molecules are given in the above-mentioned paper.

The structural parameters of Mo(CO)$_6$ and W(CO)$_6$ are very similar. Trends in small differences are, as expected, observed for the respec-

* For discussion of linear and non-linear shrinkages, see Cyvin (1968)

TABLE 44

Some parameters for the bonds in $M(CO)_6$ carbonyls (M = Cr, Mo or W)

	$r(M–C)$, Å	$l(M–C)$, Å	$r(C–O)$, Å	$l(C–O)$, Å	$k(C–O)$, mdyn/Å
CO			1.128 (a)		19.02 (e)
			1.16 ±0.05 (b)		
$Cr(CO)_6$	1.92 ±0.04 (b)				
$Mo(CO)_6$	$2.063_4 \pm 0.0030$ (c)	$0.063_0 \pm 0.0040$ (d)	$1.145_2 \pm 0.0020$ (c)	$0.034_2 \pm 0.0020$ (d)	18.22 (e)
		0.0574 (f)		0.0342 (f)	
$W(CO)_6$	$2.058_9 \pm 0.0030$ (c)	$0.058_0 \pm 0.0030$ (d)	$1.148_8 \pm 0.0025$ (c)	$0.037_5 \pm 0.0030$ (d)	17.695 (e)
		0.0534 (f)		0.0344 (f)	

(a) r_e parameter cited by Arnesen and Seip (1966).
(b) Visual technique of electron diffraction, Brockway, Ewens and Lister (1938).
(c) r_g value, electron diffraction, Seip (1967).
(d) Electron diffraction, Seip (1967).
(e) Jones (1963).
(f) Calculated from spectroscopic data by Brunvoll (1967a) using the assignment given by Jones (1963).

tive $r(C-O)$ bond lengths and force constants, including the data on *carbon monoxide* (cf., Table 44). There is no indication of any deviation from the O_h symmetry of the molecular configurations discussed, according to the electron diffraction data. This conclusion is supported by the good agreement between the electron diffraction and spectroscopic results for mean amplitudes of vibration and shrinkage effects, as seen in Table 45. The appearance of the negative shrinkage values is explained by the large uncertainties in the respective parameters, which are a result of their strong correlation (Arnesen and Seip, 1966; Seip, 1967). There is also vapour-phase vibrational spectroscopic evidence for the O_h symmetry of the molecular configurations (Hawkins, Mattraw, Sabol and Carpenter, 1955; Kawai and Murata, 1960).

The trigonal bipyramidal configuration of *iron pentacarbonyl*, $Fe(CO)_5$, (see Fig. 53) was established by the visual electron diffrac-

TABLE 45

Mean amplitudes of vibration (l, Å) and shrinkage effects (δ, Å) of non-bond distances determined by electron diffraction (ED)* and calculated from spectroscopic data (SP)** for $Mo(CO)_6$ and $W(CO)_6$

Atomic pairs[+]	l(ED)	l(SP)	δ(ED)	δ(SP)
Mo . . O1	0.056	0.0571	0.009	0.0060
C10 . . . O1	0.073	0.0734	0.010	0.0272
O1 . . . O4	0.073	0.0737	0.036	0.0397
C7 . . . C10	0.072	0.0729	0.010	0.0170
C7 . . . C8	0.131	0.1747	−0.037	0.0042
C8 . . . O1	0.218	0.2086	0.002	0.0099
O1 . . . O2	0.294	0.2737	0.012	0.0159
W . . . O1	0.058_5	0.0532	0.013	0.0060
C10 . . . O1	0.073	0.0704	0.026	0 0244
O1 . . . O4	0.092	0.0707	0.058	0.0360
C7 . . . C10	0 071	0.0700	0.002	0.0146
C7 . . . C8	0.160	0.1653	−0.027	0.0038
C8 . . . O1	0.198	0.1976	−0.009	0.0094
O1 . . . O2	0.286	0.2591	0.033	0.0154

* Seip (1967).
** Brunvoll (1967a).
[+] The numbering of atoms is given in Fig. 52.

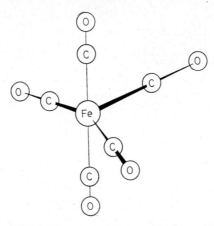

Fig. 53. The trigonal bipyramidal configuration of iron pentacarbonyl, Fe(CO)$_5$

tion investigation of Ewens and Lister (1939). The X-ray diffraction
study of Hanson (1962) showed that the molecules also had a trigonal
bipyramidal structure in the crystalline phase. The axial bonds were
found to be longer than the equatorial bonds, although the difference
was not significant (Donohue and Caron, 1964). The first sector
electron diffraction study on iron pentacarbonyl was performed by
Davis and Hanson (1965, 1967), whose data showed that the axial
bonds are actually shorter than the equatorial bonds in the vapour
phase. As is well known, the opposite trend is observed in other
trigonal bipyramidal systems. At the same time, the difference found
between the two Fe − C bond distances was near the limits of what
could be reliably established by electron diffraction. For the above
reasons, it is not surprising that the results of Davis and Hanson
generated some discussion in the literature (Donohue and Caron,
1966, 1967; Davis and Hanson, 1967).

The results of two more recent electron diffraction studies by
Beagley, Cruickshank, Pinder, Robiette and Sheldrick (1969) and
Almenningen, Haaland and Wahl (1969b) are in agreement, however,
with the findings of Davis and Hanson (1965, 1967) concerning the
Fe − C bonds in Fe(CO)$_5$. Some of the structural parameters determined by the three independent investigations are presented in Table
46. The two more recent studies are more complete than the first.

TABLE 46

Bond lengths (Å) in iron pentacarbonyl as determined by electron diffraction

	(a)	(b)	(c)
$r(C-O)$, mean	1.136 ± 0.0015	1.147 ± 0.002	1.145 ± 0.003
$r(Fe-C)$, mean	1.823 ± 0.0014	1.827 ± 0.003	1.822 ± 0.003
$\Delta r(Fe-C)$*	0.049 ± 0.020	$0.000 - 0.050$	0.027 ± 0.005

* $\Delta r(Fe-C) = r(Fe-C)_{equatorial} - r(Fe-C)_{axial}$.
(a) Davis and Hanson (1965, 1967).
(b) Almenningen, Haaland and Wahl (1969b).
(c) Beagley, Cruickshank, Pinder, Robiette and Sheldrick (1969).

As it appears from the papers on the first work, the influence of the assumptions for l values has not been checked thoroughly enough in the determination of the geometrical parameters. Also, the shrinkage effect was not taken into consideration. The later studies utilized the shrinkage values calculated by Brunvoll (1967b) from the vibrational spectroscopic data of Edgell, Wilson and Summit (1963). The work of Beagley, Cruickshank, Pinder, Robiette and Sheldrick (1969) laid much emphasis on trying various weighting schemes for the measurements. The experiments of Almenningen, Haaland and Wahl (1969b) are distinguished by the largest data interval, which is particularly valuable in resolving closely-packed internuclear distances.

As has already been mentioned, the difference observed in the axial and equatorial $Fe-C$ bonds of iron pentacarbonyl shows a trend opposite to that for similar configurations, such as those observed in PF_5 (Hansen and Bartell, 1965) or PCl_5 (Adams and Bartell, 1971). According to Davis and Hanson (1965), the reason for this deviation may be sought in the increased role of the d orbitals and the appearance of back-coordination in iron pentacarbonyl. The axial bonds were shown to be more favourable for back-coordination by molecular orbital calculations. Donohue and Caron (1967) felt that such a discussion of the bonding was premature at that time. Today, it can be stated with confidence that the axial bonds are shorter than the equatorial bonds in iron pentacarbonyl, and, as will be seen later, in other derivatives also.

A regular tetrahedral structure was found in *nickel tetracarbonyl*, $Ni(CO)_4$, by the visual electron diffraction study of Brockway and

190

Cross (1935), which has remained the only determination of the geometrical parameters to date. The following values for the bond distances were obtained: $r(Ni-C) = 1.82 \pm 0.03$ Å and $r(C-O) = 1.15 \pm 0.02$ Å. The square planar configuration was excluded, although that structural type often occurs among the complexes of nickel.

Matrix isolation infrared and laser Raman spectroscopic investigations by Kündig, Moskovits and Ozin (1972) also established regular tetrahedral configurations for $Pd(CO)_4$ and $Pt(CO)_4$.

SUBSTITUTED CARBONYLS AND PF_3 COMPLEXES

The determination of the molecular geometries of substituted carbonyls and, above all, the examination of the influence of various substituents, is of special importance in studying the bonding of transition metal carbonyls. There are few reliable data for the covalent radii of the transition metals and for the normal lengths of single and multiple bonds formed by transition metal atoms (cf., Cotton and Wilkinson, 1972; Pauling, 1960). For these reasons, it may be particularly useful to investigate the variations in the geometrical parameters caused by a series of substituents in analogous compounds. Following the early visual electron diffraction studies, there was a long break until recently when renewed interest was shown and a few up-to-date investigations were carried out. It will, however, probably require the collection of data for several more years in order to provide the possibility of establishing correlations of general value between the bonding and molecular geometries of carbonyl complexes.

Trifluorophosphine as a ligand has special importance among substituted carbonyls, as was noted by Parry (1969) and Nixon (1973), for example. In order to supplement the data on their electronic structure discussed in the literature (see, *e.g.*, Nixon, 1973), geometrical evidence is presented below to show the analogous behaviour of the carbonyl and trifluorophosphine ligands in transition metal complexes. Thus, a discussion of the complexes containing only trifluorophosphine groups as ligands together with the carbonyl derivatives can be considered to be adequate.

Among the small number of studies, a high proportion dealt with manganese compounds. Their structural parameters are given in Table 47.

191

TABLE 47

Geometrical parameters for substituted manganese carbonyls

	HMn(CO)₅ (a)	HMn(CO)₅ (b)	CH₃Mn(CO)₅ (c)	Br₃GeMn(CO)₅ (d)	(CO)₅MnMn(CO)₅ (e)	(CO)₅MnMn(CO)₅ (f)	(CO)₅MnMn(CO)₅ (g)
r(Mn—X), Å	1.50±0.07	1.601±0.016	2.185±0.011	2.44	2.91±0.02	2.977±0.011	2.923±0.003
r(Mn—C)$_{mean}$, Å	1.860±0.004						
r(Mn—C)$_{eq.}$, Å		1.853±0.012	1.860±0.004	1.84	1.86±0.01	1.873±0.005	1.830±0.008
r(Mn—C)$_{ax.}$, Å		1.822±0.012	1.820 (assumed)	1.79	1.82±0.02	1.803±0.016	1.792±0.014
r(C—O), Å	1.139±0.004	1.134±0.004	1.141±0.002	1.16	1.12±0.01	1.147±0.002	1.156±0.007
$\angle$C$_{eq.}$—Mn—C$_{ax.}$	96.4±1.0°	94.9—98.2°	94.7±1.0°	95°		93.4±0.5°	93.8±0.4°

(a), (c), (d), (e) and (f) Vapour-phase electron diffraction studies.

(a) Robiette, Sheldrick and Simpson (1969).

(b) Crystal-phase X-ray and neutron diffraction investigation, La Placa, Hamilton, Ibers and Davison (1969).

(c) Seip and Seip (1970).

(d) The uncertainties of the parameters are not given; Gapotchenko, Alekseev, Antonova, Anisimov, Kolobova, Ronova and Struchkov (1970).

(e) Gapotchenko, Alekseev, Anisimov, Kolobova and Ronova (1968).

(f) Almenningen, Jacobsen and Seip (1969).

(g) Crystal-phase X-ray diffraction, Dahl and Rundle (1963).

The electron diffraction structural analysis of *manganese penta-carbonyl hydride* was performed by assuming a molecular model with C_{4v} symmetry (Robiette, Sheldrick and Simpson, 1968, 1969). This assumption is in agreement with the results of crystal-phase structural determinations (La Placa, Hamilton, Ibers and Davison, 1969). The values for $r(Mn-C)$ and $r(C-O)$ determined in the crystalline and gaseous phases are also consistent. The earlier reported electron diffraction value for $r(Mn-H) = 1.426 \pm 0.046$ Å was later found to be in error and the corrected value, 1.50 ± 0.07 Å, although uncertain, does not contradict the crystal-phase diffraction data. It is hardly probable that the values of $r(Mn-H)$ would differ considerably in the two phases. The hydrogen atom was located with a precision of 0.01 Å in the study of La Placa, Hamilton, Ibers and Davison (1969). This was made possible by employing the neutron diffraction technique, as the scattering amplitude of neutrons on hydrogen is nearly equal to that on manganese and is only slightly less than those on carbon and oxygen. The value of the bond distance $r(Mn-H)$ in $HMn(CO)_5$ indicates the presence of a normal covalent bond. The predicted value for the length of the single $Mn-H$ bond was 1.65 Å (Handy, Treichel, Dahl and Hayter, 1966) based on the estimated values for the covalent radii of manganese and hydrogen.

The determination of the covalent radius of the manganese atom was the main purpose of the electron diffraction study on *methyl-manganese pentacarbonyl*, $CH_3Mn(CO)_5$, by Seip and Seip (1970). Using the value of 0.767 Å for the carbon covalent radius, the covalent radius of manganese was determined to be 1.41 Å, in agreement with earlier estimates (Cotton and Richardson, 1966).

Preliminary results have been published only for the molecular geometry of *(tribromogermyl) manganese pentacarbonyl*, $Br_3GeMn(CO)_5$ by Gapotchenko, Alekseev, Antonova, Anisimov, Kolobova, Ronova and Struchkov (1970). The parameters characterizing the bond configuration of the $GeBr_3$ group are given in Fig. 54.

For the sake of comparison, Table 47 also contains structural parameters for dimanganese decacarbonyl, $Mn_2(CO)_{10}$, in addition to the data for substituted manganese carbonyls. Binuclear carbonyls are dealt with in more detail in a subsequent section.

Two striking peculiarities can be observed in the molecular geometries of the substituted manganese carbonyls: (1) the equatorial

13   193

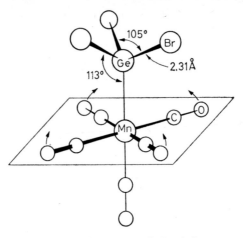

Fig. 54. The configuration of the (tribromogermyl)manganese pentacarbonyl, $Br_3GeMn(CO)_5$, molecule and the structural parameters of the $GeBr_3$ group (Gapotchenko, Alekseev, Antonova, Anisimov, Kolobova, Ronova and Struchkov, 1970). The deviation of the carbonyl groups from the equatorial plane is indicated by arrows

manganese–carbon bonds are longer than the axial bonds,* and (2) the equatorial Mn−C−O groups deviate from the equatorial plane towards the substituent. This deviation is indicated by arrows in Fig. 54. The difference between the equatorial and axial bonds is of the same sign as observed in iron pentacarbonyl. The problems in the interpretation of this phenomenon have already been indicated. The deviation of the equatorial manganese carbonyl groups is consistent with the general observation that the metal carbonyl groups always deviate *from* the stronger π-acceptor ligand (cf., Cotton and Wilkinson, 1972).

The following substituted iron and cobalt carbonyls have been studied by the visual technique of electron diffraction:

$$\left.\begin{array}{l} FeH_2(CO)_4 \\ CoH(CO)_4 \end{array}\right\} \text{ Ewens and Lister (1939),}$$

---

* It is a misprint in the abstract in the paper by La Placa, Hamilton, Ibers and Davison (1969) that the axial bond is indicated to be longer.

$$\left.\begin{array}{l} \text{Fe(CO)}_2\text{(NO)}_2 \\ \text{Co(CO)}_3\text{(NO)} \end{array}\right\} \text{ Brockway and Anderson (1937).}$$

The nitrosyl derivatives have a tetrahedral configuration and the bond distances are given in Table 48. The metal–carbon and metal–

TABLE 48

Bond distances in substituted iron and cobalt carbonyls as determined by the visual technique of electron diffraction by Brockway and Anderson (1937)

	$\text{Fe(CO)}_2\text{(NO)}_2$	$\text{Co(CO)}_3\text{(NO)}$
$r(M-C)$, Å	1.84 ± 0.02	1.83 ± 0.02
$r(M-N)$, Å	1.77 ± 0.02	1.76 ± 0.03
$r(C-O)$, Å	1.15 ± 0.03	1.14 ± 0.03
$r(N-O)$, Å	1.12 ± 0.03	1.10 ± 0.04

nitrogen bonds are all shorter than the sums of the respective covalent radii corresponding to single bonds. The data are much too uncertain to justify a detailed discussion. The conclusions of the early electron diffraction investigation concerning the structures of the hydrides (*viz.*, that the molecules are tetrahedral and the hydrogen atoms are linked to the oxygen atoms of the CO groups) proved to be in error. Later studies showed the presence of direct metal–hydrogen bonds in these molecules (for more detail see, *e.g.*, Bor, 1966; Cotton and Wilkinson, 1972; Pauling, 1960).

The vapour-phase molecular geometry of *silylcobalt tetracarbonyl*, $H_3SiCo(CO)_4$, was determined by the electron diffraction study of Robiette, Sheldrick, Simpson, Aylett and Campbell (1968). The structural analysis was performed for a model with C_{3v} symmetry shown in Fig. 55. The two types of $Co-C$ and $C-O$ bonds were each treated as having equal lengths. Assuming the value of 1.48 Å for $r(Si-H)$, the following bond distances and bond angle were obtained:

$$
\begin{aligned}
r(C-O) &= 1.137 \pm 0.003 \text{ Å} \\
r(Co-C) &= 1.802 \pm 0.003 \text{ Å} \\
r(Co-Si) &= 2.381 \pm 0.007 \text{ Å} \\
<Si-Co-C &= 81.7 \pm 0.7°.
\end{aligned}
$$

13*

The cobalt–silicon bond in the vapour-phase molecule was found to be considerably longer than that observed in crystalline $Cl_3SiCo(CO)_4$, i.e., 2.254 ± 0.003 Å (Robinson and Ibers, 1967). The other parameters appearing in both structures are in good agreement. The shortening of the Co−Si bond is interpreted as being a consequence of the substitution of the electron-withdrawing groups on the silicon atom. The presence of these groups decreases the energy of the d orbitals and, accordingly, increases their involvement in $d \rightarrow d$ π-bonding. As stated in the authors' discussion, it is less probable that the change is due to the electronegativity differences. The lengths of the Si−H and C−H bonds have been shown to be insensitive to a change in the electronegativity of the other substituents on the silicon or carbon atoms (Ebsworth, 1963). Again the deviation of the equatorial carbonyl groups corresponds to the general tendency of deviation *from* the stronger π-acceptor ligand.

It should be mentioned, in connection with the electron diffraction study of $H_3SiCo(CO)_4$, that most of the mean amplitudes of vibration used in the structural analysis were assumed and not refined. The four l values determined are given below together with some calculated data from the spectroscopic study of Devarajan and Cyvin (1972) on $Cl_3SiCo(CO)_4$:

	$H_3SiCo(CO)_4$, measured	$Cl_3SiCo(CO)_4$, calculated
$l(C-O)$, Å	0.044 ± 0.009	0.0342
$l(Co-C)_{ax.}$, Å		0.0486
$l(Co-C)_{eq.}$, Å	0.057 ± 0.005	0.0543
$l(Co-Si)$, Å	0.078 ± 0.007	0.0602
$l(Co \ldots O)_{ax.}$, Å		0.0497
$l(Co \ldots O)_{eq.}$, Å	0.065 ± 0.004	0.0546

Spectroscopic calculations were also carried out for $Cl_3GeCo(CO)_4$ and $Cl_3SnCo(CO)_4$ (Devarajan and Cyvin, 1972).

The geometrical data characterizing the bond configuration of the PF_3 group in *(trifluorophosphine)molybdenum pentacarbonyl*, $PF_3Mo(CO)_5$, are listed in Table 49 together with the parameters for other trifluorophosphine derivatives. The bond distances and bond angles for the remainder of the $PF_3Mo(CO)_5$ molecule are shown in Fig. 56. All these are $r_g(1)$ values obtained in an electron

196

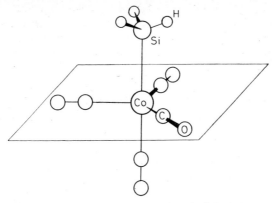

Fig. 55. The molecular model of silylcobalt tetracarbonyl, $H_3SiCo(CO)_4$

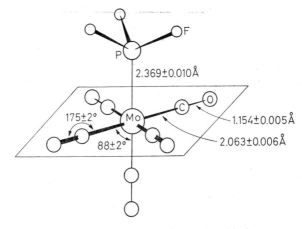

Fig. 56. The configuration of the (trifluorophosphine)molybdenum pentacarbonyl molecule, $PF_3Mo(CO)_5$, and the structural parameters of the $Mo(CO)_5$ group (Bridges, Holywell, Rankin and Freeman, 1971)

diffraction investigation by Bridges, Holywell, Rankin and Freeman (1971). The structural parameters for the molybdenum pentacarbonyl part, including the mean amplitudes of vibration, are in good agreement with the data for molybdenum hexacarbonyl (Arnesen and Seip, 1966).

Two independent electron diffraction studies were performed at about the same time in order to elucidate the molecular structure of *tetrakis(trifluorophosphine)nickel*, $Ni(PF_3)_4$ (Almenningen, Andersen and Astrup, 1970; Marriott, Salthouse and Ware, 1970). Almenningen, Andersen and Astrup (1970) gave a detailed description of the structural analysis that established the free rotation of the PF_3 groups around the nickel–phosphorus axis. It was found that the nickel atom has a tetrahedral bond configuration, as shown in Fig. 57, while the square planar arrangement

could be excluded with confidence. The geometrical parameters for the phosphorus bond configuration are given in Table 49, together with those for *tetrakis(trifluorophosphine)platinum*, $Pt(PF_3)_4$, determined by Marriott, Salthouse and Ware (1970).

There is excellent agreement between the geometrical parameters determined in the two electron diffraction studies on $Ni(PF_3)_4$. On the other hand, there are considerable differences concerning the mean amplitudes of vibration. The l values associated with the rotation-dependent distances are particularly uncertain. In this connection, it is interesting to note that the l values for the non-bond distances change drastically with temperature, according to the findings of a spectroscopic study by Cyvin and Müller (1971).

The nickel–phosphorus bond in $Ni(PF_3)_4$ is the shortest ever observed in nickel–phosphorus complexes (Nixon, 1973; Mais, Owston, Thompson and Wood, 1967; Jarvis, Mais and Owston, 1968). The molybdenum–phosphorus bond in $PF_3Mo(CO)_5$ is also considerably shorter than the Mo – P bonds found in other molybdenum–phosphine

198

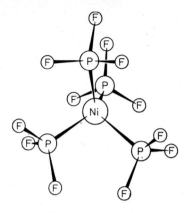

Fig. 57. The tetrahedral structure of
tetrakis(trifluorophosphine)nickel, Ni(PF$_3$)$_4$
(Almenningen, Andersen and Astrup, 1970)

TABLE 49

Bond configurations of the PF$_3$ groups in trifluorophosphine complexes

		$r(P-F)$, Å	$<F-P-F$	$r(P-M)$, Å
PF$_3$	(a)	1.569 ± 0.001	97.7 ± 0.2°	
(PF$_3$)$_4$Ni	(b)	1.561 ± 0.003	99.3 ± 0.3°	2.099 ± 0.003
	(c)	1.561 ± 0.005	98.4 ± 0.8°	2.116 ± 0.010
PF$_3$Mo(CO)$_5$	(d)	1.557 ± 0.004	99.5 ± 0.5°	2.369 ± 0.010
(PF$_3$)$_4$Pt	(c)	1.546 ± 0.006	98.9 ± 0.7°	2.230 ± 0.010
PF$_3$BH$_3$	(e)	1.538 ± 0.008	99.8 ± 1.0°	1.836 ± 0.012

(a), (b), (c) and (d) Electron diffraction, $r_g(1)$ parameters.
(a) Morino, Kuchitsu and Moritani (1969).
(b) Almenningen, Andersen and Astrup (1970).
(c) Marriott, Salthouse and Ware (1970).
(d) Bridges, Holywell, Rankin and Freeman (1971).
(e) Microwave spectroscopy, r_0 parameters, Kuczkowski and Lide (1967).

complexes (Mais, Owston and Thompson, 1967; Payne, Mokuolu
and Speakman, 1965). The platinum–phosphorus bond in Pt(PF$_3$)$_4$ is,
however, only slightly shorter than is usually observed (Eisenberg
and Ibers, 1965; Messmer and Amma, 1966). The variations of the
metal–phosphorus bond lengths are correlated with the changes in the

199

electronegativity of the substituent in the PX$_3$ group. As discussed by Bridges, Holywell, Rankin and Freeman (1971), the replacement of the alkyl or aryl groups with fluorine atoms in the phosphine ligand increases greatly the π-acceptor properties of the ligand.

The PF$_3$ ligand is a π-acceptor with a strength commensurate with that of the carbonyl ligand. This is strongly supported by the following geometrical evidence:

(i) The equatorial carbonyl groups do not deviate from the equatorial plane within the experimental error limit.

(ii) The difference between the bond distances $r(\text{Mo}-\text{C})$ and $r(\text{Mo}-\text{P})$ in PF$_3$Mo(CO)$_5$ is the same (0.306 Å) as that between the bond distances $r(\text{C}-\text{C})$ of ethane and $r(\text{P}-\text{C})$ of trimethylphosphine (0.307 Å), as pointed out by Bridges, Holywell, Rankin and Freeman (1971).

It is also of interest to follow the variations in the bond configurations of the PF$_3$ group itself. The data necessary for such a comparison are summarized in Table 49. The length of the phosphorus–fluorine bond in all of the complexes is shorter than that in the free molecule. Also, the F$-$P$-$F bond angles are larger in the complexes. These variations are probably correlated with changes in the strength of the metal–phosphorus (M$-$P) bonds. The bond distance $r(\text{P}-\text{F})$ in Ni(PF$_3$)$_4$ is hardly distinguishable from that in free PF$_3$, in agreement with the observation of the extremely short Ni$-$P bond. Not considering the addition compounds, the greatest shortening of the P$-$F bond occurs in the platinum derivative where the Pt$-$P bond is of ordinary strength. The parameters for the molybdenum compound indicate an intermediate position. The variations in the bond lengths, as demonstrated above, can easily be accounted for by the concept that the extent of the back-coordination from the metal atom towards the ligand gradually decreases in the order Ni > Mo > > Pt. These observations are then well supplemented by the fact that the PF$_3$.BH$_3$ complex shows the largest deviation of the PF$_3$ geometry in comparison with the free phosphine molecule. In the boron–phosphorus addition compound, of course, no back-coordination is expected to occur.

The most interesting feature of the structures of the olefin complexes is the intermediate character between the σ- and π-bonding of the metal to organic group linkage (cf., Cotton and Wilkinson, 1972). The vapour-phase molecular geometries of five iron carbonyl derivatives plus that of π-allyl cobalt tricarbonyl have been elucidated so far.

The molecular models were assumed to possess C_{2v} symmetry in the electron diffraction studies on both *ethylene iron tetracarbonyl*, $C_2H_4Fe(CO)_4$, by Davis and Speed (1970) and *tetrafluoroethylene iron tetracarbonyl*, $C_2F_4Fe(CO)_4$, by Beagley, Schmidling and Cruickshank (1973). The iron atom has a distorted octahedral bond configuration in such a model. The two carbon atoms of the ethylene group, two of the carbonyl groups and the iron atom itself are situated in the equatorial plane. The molecular configurations as described here are supported by ^{19}F NMR and infrared spectroscopic evidence (Fields, Germain, Haszeldine and Wiggans, 1970). The bond configuration of the iron atom can be considered to be octahedral, of course, only when σ-bonds are formed. In the C_2F_4 derivative, the coordination linkage is indeed σ rather than π in character, according to Beagley, Schmidling and Cruickshank (1973). On the other hand, the iron atom bond configuration can be considered to be trigonal bipyramidal in ethylene iron tetracarbonyl, as the metal–olefin linkage has more π-character. The ethylene molecule is then occupying one of the equatorial positions. The molecular models of both compounds mentioned are shown in Fig. 58. The deviation of the CF_2 groups from the metal atom is also indicated. An analogous distortion of the olefin group was not examined in $C_2H_4Fe(CO)_4$ because of the weak electron scattering properties of the hydrogen atom. The geometrical parameters of both molecules are summarized in Table 50.

In connection with these studies, a final remark is felt to be necessary. The physical techniques, including the electron diffraction method, that are used in order to determine the relative configurations of the atomic nuclei of a molecule are not capable of providing more direct evidence for the existence of a chemical bond. Thus, for example, it cannot be decided from the geometrical parameters alone whether the structures discussed above should be considered as distorted octahedral or trigonal bipyramidal configurations.

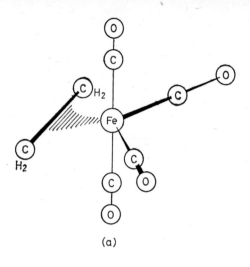

(a)

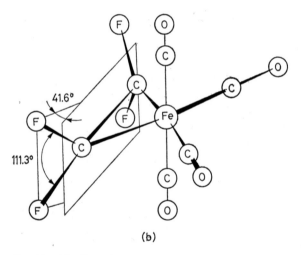

(b)

Fig. 58. (a) The trigonal bipyramidal configuration of ethylene iron tetracarbonyl, $C_2H_4Fe(CO)_4$; (b) the distorted octahedral configuration of tetrafluoroethylene iron tetracarbonyl, $C_2F_4Fe(CO)_4$ (Beagley, Schmidling and Cruickshank 1973)

TABLE 50

Geometrical parameters for ethylene iron tetracarbonyl, $C_2H_4Fe(CO)_4$, and tetrafluoroethylene iron tetracarbonyl, $C_2F_4Fe(CO)_4$, as determined by electron diffraction

	$C_2H_4Fe(CO)_4$ (a)	$C_2F_4Fe(CO)_4$ (b)
$r(Fe-CO)_{eq.}$, Å	1.84 ± 0.02	1.846 ± 0.010
$r(Fe-CO)_{ax.}$, Å	1.80 ± 0.02	1.823 ± 0.010
$r(Fe-CO)_{mean}$, Å	1.82	1.835 ± 0.006
$r(Fe-C)$, Å	2.12 ± 0.02	1.989 ± 0.010
$r(C-C)$, Å	1.46 ± 0.03	1.530 ± 0.015
$<C_{eq.}-Fe-C_{eq.}$	$105 \pm 2°$	$104.2 \pm 1.4°$

(*a*) Davis and Speed (1970); the values and uncertainties are given according to Beagley, Schmidling and Cruickshank (1973).

(*b*) Beagley Schmidling and Cruickshank (1973).

It is not easy to judge the reliability of the structural parameters obtained from the electron diffraction study on *butadiene iron tricarbonyl*, $C_4H_6Fe(CO)_3$ (Davis and Speed, 1970). Not only was a rather simplified molecular model assumed but also many of the various assumptions were not tested for their influence on the other parameters determined, or at least no such tests were described in the paper. The overall molecular model used is shown in Fig. 59. Some of the assumptions were, *e.g.*, that the $Fe(CO)_3$ group had C_{3v} symmetry and all carbon–carbon bonds in the butadiene part were of equal length. On the other hand, a detailed analysis was carried out for determining some parameters, definitely of lesser importance. The more interesting bond distances were given as follows:

$$r(Fe-CO) = 1.798 \pm 0.015 \text{ Å}$$
$$r(Fe-C) = 2.074 \pm 0.015 \text{ Å}$$
$$r(C-C) = 1.413 \pm 0.015 \text{ Å}$$
$$r(C-O) = 1.137 \pm 0.010 \text{ Å}.$$

The $OC-Fe-CO$ bond angle was found to be $100.5 \pm 5.0°$.

Cyclobutadiene iron tricarbonyl, $C_4H_4Fe(CO)_3$, was the subject of two independent electron diffraction investigations performed at about the same time (Davis and Speed, 1970; Oberhammer and Brune, 1969). The results are summarized in Table 51. The differences

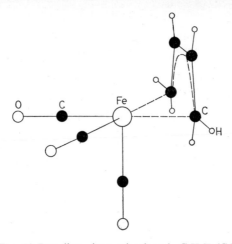

Fig. 59 Butadiene iron tricarbonyl, $C_4H_6Fe(CO)_3$
(Davis and Speed, 1970)

TABLE 51

Geometrical parameters for cyclobutadiene iron tricarbonyl, $C_4H_4Fe(CO)_3$, and tetraphenyl cyclobutadiene iron tricarbonyl, $(C_6H_5)_4C_4Fe(CO)_3$

		$C_4H_4Fe(CO)_3$		$(C_6H_5)_4C_4Fe(CO)_3$
		(a)	(b)	(c)
$r(Fe-CO)$,	Å	1.819 ± 0.010	1.778 ± 0.004	1.750
$r(Fe-C)$,	Å	2.063 ± 0.010	2.051 ± 0.004	2.065
$r(C-C)$,	Å	1.456 ± 0.015	1.439 ± 0.006	1.459
$r(C-O)$,	Å	1.131 ± 0.010	1.146 ± 0.003	1.179
$<OC-Fe-CO$		$95.5\pm2.5°$	$98.1\pm0.8°$	$97.0°$
δ		—	$8.6\pm6.3°$	—

(a) Electron diffraction, Davis and Speed (1970).

(b) Electron diffraction, $r_g(1)$ parameters; δ is the angle by which the C—H bond deviates from the plane of the cyclobutadiene ring towards the iron atom; Oberhammer and Brune (1969).

(c) Crystal-phase X-ray diffraction, mean values of the parameters; Dodge and Schomaker (1965).

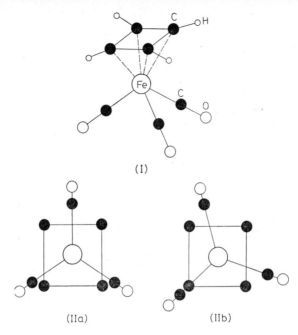

(I)

(IIa)   (IIb)

Fig. 60. Cyclobutadiene iron tricarbonyl, $C_4H_4Fe(CO)_3$ (Oberhammer and Brune, 1969; Davis and Speed, 1970). (I) The molecular configuration; (IIa) and (IIb) two relative orientations of the cyclobutadiene ring and the $Fe(CO)_3$ group

between the data in the two papers are considerable for some of the parameters. The molecular model is shown in Fig. 60. The relative orientation of the cyclobutadiene ring and the $Fe(CO)_3$ fragment could not be determined unambiguously. The possibility of free rotation also could not be ruled out. For the sake of comparison, the structural parameters determined for the crystal·phase tetraphenyl cyclobutadiene iron tricarbonyl, $(C_6H_5)_4C_4Fe(CO)_3$, are also included in Table 51. The conformation of this molecule was found to be analogous to the (b) form in Fig. 60.

The molecular geometry of *trismethylenemethane iron tricarbonyl*, $(CH_2)_3CFe(CO)_3$, was found to possess C_{3v} symmetry with a staggered arrangement of the ligands (see Fig. 61) in the electron diffraction investigation by Almenningen, Haaland and Wahl (1969a). In addi-

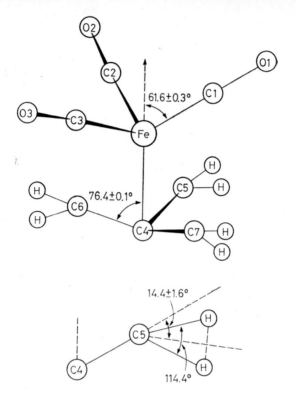

Fig. 61. The molecular model of trismethylenemethane iron tricarbonyl, $(CH_2)_3CFe(CO)_3$, with the angular parameters (Almenningen, Haaland and Wahl, 1969a)

tion, the C_{3v} symmetry was also established from an earlier proton magnetic resonance spectrum consisting of a sharp singlet, thus indicating that all hydrogen atoms are equivalent (Emerson, Ehrlich, Giering and Lauterbur, 1966). The following bond distances were determined by the electron diffraction study:

$$r(Fe-CO) = 1.810 \pm 0.003 \text{ Å}$$
$$r(Fe-C) = 1.938 \pm 0.004 \text{ Å}$$
$$r(C-C) = 1.437 \pm 0.003 \text{ Å}$$
$$r(C-O) = 1.111 \pm 0.007 \text{ Å}.$$

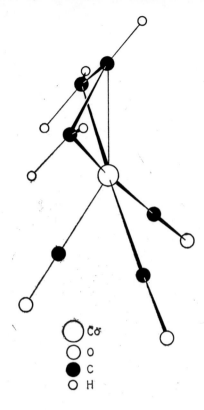

Fig. 62. π-Allyl cobalt tricarbonyl, $C_3H_5Co(CO)_3$
(Seip, 1972)

The angular parameters are presented in Fig. 61. The peculiarity of the structure is that the CH_2 groups in the trismethylenemethane fragment are bent towards the metal atom. The mean amplitudes of vibration have been studied in detail. The molecule is relatively rigid, as shown by the small *l* values associated with atomic pairs in which one of the atoms belongs to the $(CH_2)_3C$ group and the other to the carbonyl part. Accordingly, the barrier to internal rotation around the iron–carbon(methane) linkage was expected to be high.

The electron diffraction data of *π-allyl cobalt tricarbonyl*, $C_3H_5Co(CO)_3$, were consistent with a model having C_s symmetry in which the cobalt atom is six-coordinated, as shown in Fig. 62. The following bond lengths and bond angles were determined (Seip, 1972):

$r(\text{Co}-\text{CO}) = 1.804 \pm 0.003 \text{ Å}$ $<\text{OC}-\text{Co}-\text{CO} = 100.3 \pm 1.0°$
$r(\text{Co}-\text{C}) = 1.985 \pm 0.016 \text{ Å}$ $<\text{OC}-\text{Co}-\text{CO} = 109.0 \pm 2.2°$
$r(\text{Co}-\text{C}) = 2.101 \pm 0.006 \text{ Å}$ $<\text{C}-\text{C}-\text{C} = 123.2 \pm 3.1°$
$r(\text{C}-\text{C}) = 1.391 \pm 0.009 \text{ Å}$
$r(\text{C}-\text{O}) = 1.144 \pm 0.003 \text{ Å}$

Several interesting conclusions have already been drawn by different authors from the geometrical data concerning bonding in the olefin complexes of transition metals. More systematic and extended studies have yet to appear on the detailed analysis of the correlation between the molecular geometry and bonding. The most direct information on the bond strength is provided, of course, by the value of the distance between the iron (or cobalt) atom and the olefin group. Thus the iron–olefin linkage is stronger in $C_2F_4Fe(CO)_4$ than in $C_2H_4Fe(CO)_4$. The strengthening of the iron–olefin bond is accompanied by a weakening (i.e., lengthening) of the C−C bond (cf., Table 50). The difference between the equatorial and axial Fe−CO bonds in these complexes shows the same trend although it is less pronounced than that in $Fe(CO)_5$ (see Table 46). A wider variation is observed for the values of $r(\text{Fe}-\text{C})$ for the iron–olefin than for the iron–carbonyl linkage. However, no correlation can be established on the basis of available data, although a weakening of the iron–carbonyl bond might be expected as the iron–olefin bond becomes stronger.

Finally, comparison of the geometrical data for complexed and free ethylene provides some information of interest. The carbon–carbon bond in $C_2H_4Fe(CO)_4$ is considerably longer (1.46 Å) than that in ethylene, $1.337 \pm 0.002 \text{ Å}$ (Bartell, Roth, Hollowell, Kuchitsu and Young, 1965; Kuchitsu, 1966). This bond is even longer in $C_2F_4Fe(CO)_4$ (1.530 Å), where the CF_2 groups were found to bend away from the iron atom (see Fig. 58). Correlation is believed to exist between the bond lengthening and the distortion of the configuration (Beagley, Schmidling and Cruickshank, 1973; Stalick and Ibers, 1970; McAdam, Francis and Ibers, 1971a, 1971b).

The geometrical parameters for *dimanganese decacarbonyl*, $Mn_2(CO)_{10}$, originating from different studies, are given in Table 47. Spectroscopically calculated mean amplitudes of vibration obtained by Brunvoll and Cyvin (1968) were utilized in an electron diffraction structural analysis by Almenningen, Jacobsen and Seip (1969). As indicated in the latter paper, further refinement of the electron diffraction data was planned when more vibrational spectroscopic information (Adams and Squire, 1968) became available. The two most important features of the $Mn_2(CO)_{10}$ molecular geometry are the staggered orientation of the equatorial carbonyl groups belonging to different manganese atoms, as shown in Fig. 63, and the length of the Mn−Mn bond given in Table 47.

Gapotchenko, Struchkov, Alekseev and Ronova (1973) reported on the electron diffraction structural analysis of *dirhenium decacar-*

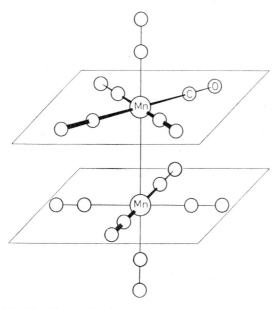

Fig. 63. The molecular configuration of dimanganese decacarbonyl, $Mn_2(CO)_{10}$. The geometrical parameters are given in Table 47

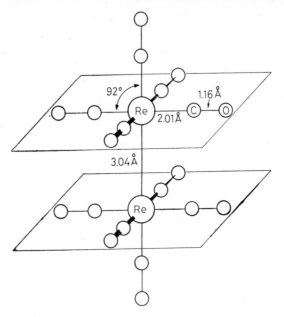

Fig. 64. The molecular configuration of dirhenium decacarbonyl, $Re_2(CO)_{10}$, and the geometrical parameters
determined by Gapotchenko, Struchkov, Alekseev and
Ronova (1973)

bonyl, $Re_2(CO)_{10}$. The relative orientation of the equatorial carbonyl
groups belonging to different rhenium atoms is eclipsed, as illustrated
in Fig. 64 (where the geometrical parameters are also shown).

The lengths of the metal–metal bonds in $Mn_2(CO)_{10}$ and $Re_2(CO)_{10}$
are considerably different, in agreement with the data determined
for bond energies (Svec and Junk, 1967). Although different values
have been reported for $r(Mn-Mn)$ by different authors (cf., Table 47),
all of them are larger than twice the covalent radius (2.76 Å). On the
other hand, the length of the $Re-Re$ bond was found to be about
the same as twice the covalent radius of 1.53 Å (Andrianov, Biryukov
and Struchkov, 1969) in both vapour- and crystal-phase studies (the
latter by Dahl, Ishishi and Rundle, 1957).

The variations in the metal–carbon and carbon–oxygen bond distances in both the unsubstituted and substituted carbonyl complexes
of transition metals show tendencies that correspond to the expecta-

tions. The value of $r(C-O)$ in all instances is larger than that in the free carbon monoxide molecule, $r_e = 1.128$ Å (as cited by Arnesen and Seip, 1966). Thus a decrease in the bond order of the $C-O$ bond is indicated as the coordination linkage is formed. The decrease in the value of $r(M-C)$ compared with the sum of the respective covalent radii suggests some π-character of the coordination bond, *i.e.*, the presence of back-coordination.

METALLOCENES

(CYCLOPENTADIENYL COMPLEXES)

There are many considerations that might serve as a basis for the classification of the cyclopentadienyl complexes. Such schemes of classification may be to divide them into the two groups of non-transition and transition metal complexes, or to separate the dicyclo-pentadienyl compounds from the so-called half-sandwiches with only one C_5H_5 ring, etc. We found that the most convenient procedure also seemed to be the simplest. We chose to deal with these compounds according to the position of the central atom in the Periodic System, considering group after group. The cyclopentadienyl compounds with the so-called classical structure, such as $C_5H_5SiH_3$

will be excluded from the discussion. A list of such compounds whose vapour-phase molecular geometries have been determined so far is as follows:

$C_5H_5SiH_3$	Bentham and Rankin (1971)
$C_5H_5Si(CH_3)_3$	Veniaminov, Ustynyuk, Alekseev, Ronova and Struchkov (1970)

$C_5H_4[Si(CH_3)_3]_2$ Veniaminov, Ustynyuk, Alekseev, Ronova and Struchkov (1972)

$C_5H_5Ge(CH_3)_3$ Veniaminov, Ustynyuk, Struchkov, Alekseev and Ronova (1970)

$C_5H_5Sn(CH_3)_3$ Veniaminov, Ustynyuk, Alekseev, Ronova and Struchkov (1971a)

$C_5H_4[Sn(CH_3)_3]_2$ Veniaminov, Ustynyuk, Alekseev, Ronova and Struchkov (1971b).

BERYLLIUM COMPOUNDS

The molecular structures of *bis(cyclopentadienyl)beryllium*, $(C_5H_5)_2Be$, and *cyclopentadienylberyllium borohydride*, $C_5H_5BeBH_4$, have already been mentioned (pp. 15 and 104). In addition to these two compounds, the following have been studied, all of them by the Oslo electron diffraction group (the references are given in Table 52): *methyl(cyclopentadienyl)beryllium*, $CH_3BeC_5H_5$, *cyclopentadienylberyllium acetylide*, C_5H_5BeCCH, *cyclopentadienylberyllium chloride*, C_5H_5BeCl, and *cyclopentadienylberyllium bromide*, C_5H_5BeBr. The molecular geometry of the chloride derivative has also been determined by microwave spectroscopy (Bjørseth, Drew, Marstokk and Møllendal, 1972).

The following assumptions have been utilized in the structural analyses: (1) the cyclopentadienyl rings have D_{5h} symmetry, *i.e.*, the π-electron density is symmetrically delocalized (the ring is of *pentahapto* type), and (2) the molecules as a whole have C_{5v} symmetry, *i.e.*, the beryllium atom lies on the five-fold symmetry axis of the ring. These assumptions were supported by vibrational spectroscopic evidence (Drew and Morgan, 1971) and confirmed by the electron diffraction data. For example, the sharpness of the $C-C$ bond distance maximum at 1.42 Å and that of the $C \ldots C$ non-bond distance at 2.30 Å, on the experimental radial distribution curve of C_5H_5BeCl, indicate that all carbon atoms occupy equivalent positions in the ring. Similar observations can be made for the distances that characterize the positions of the beryllium and chlorine atoms (Drew and Haaland, 1972d). Such observations are then strengthened on a more quantitative basis provided by the mean amplitudes of vibration for the corresponding distances obtained from the least-squares refinements.

212

TABLE 52

Bond distances (Å) in vapour-phase beryllium compounds as determined by electron diffraction

		$r(Be-C)^*_{cp}$	$r(Be-X)$	$r(C-C)^*_{cp}$
$C_5H_5BeC_5H_5$	(a)	1.907±0.005 2.256±0.007		1.425±0.002
$C_5H_5BeCH_3$	(b)	1.923±0.003	1.706±0.003	1.420±0.001
$Be(CH_3)_2$	(c)		1.698±0.005	
C_5H_5BeCCH	(d)	1.919±0.005	1.634±0.008	1.428±0.002
C_5H_5BeCl	(e)	1.916±0.006	1.837±0.006	1.424±0.001
$BeCl_2$	(f)		1.75 ±0.02	
C_5H_5BeBr	(d)	1.950±0.012	1.943±0.015	1.424±0.002
$BeBr_2$	(f)		1.92 ±0.02	
$C_5H_5BeBH_4$	(g)	1.915±0.005	1.88 ±0.01	1.422±0.001
$Be(BH_4)_2$	(h)		1.790±0.015	

* $r(Be-C)_{cp}$ = the distances between the beryllium atom and the carbon atoms of the cyclopentadienyl (cp) ring; $r(C-C)_{cp}$ = the cyclopentadienyl intra-ring bond distance.

(a) Haaland (1968).
(b) Drew and Haaland (1972c).
(c) Almenningen, Haaland and Morgan (1969).
(d) Haaland and Novak (1974).
(e) Drew and Haaland (1972d).
(f) Akishin and Spiridonov (1957).
(g) Drew, Gundersen and Haaland (1972).
(h) Gundersen, Hedberg and Hedberg (1973).

Returning to the assumptions utilized in the course of the structural analyses, the following specific constraints were applied in the study of methyl(cyclopentadienyl)beryllium: the methyl group had C_{3v} symmetry and its three-fold axis coincided with the five-fold axis of the remainder of the molecule. The refinements of the structural parameters were carried out for a molecular model with C_s symmetry in which one of the methyl hydrogen atoms eclipsed one of the hydrogen atoms of the C_5H_5 group, although it was realized that the barrier to internal rotation is probably very low. The molecular models of the C_5H_5BeX (X = Cl, Br, CH_3 or C≡CH) compounds are shown in Fig. 65, and the most important geometrical parameters are compiled in Table 52. It should be stressed that there is good agreement between the electron diffraction and microwave spectroscopic results for C_5H_5BeCl, as demonstrated by the data in Table 53. In addition to the r_0 parameters, Bjørseth, Drew, Marstokk and Møllendal (1972) also determined the substitution coordinates (r_s structure) for the

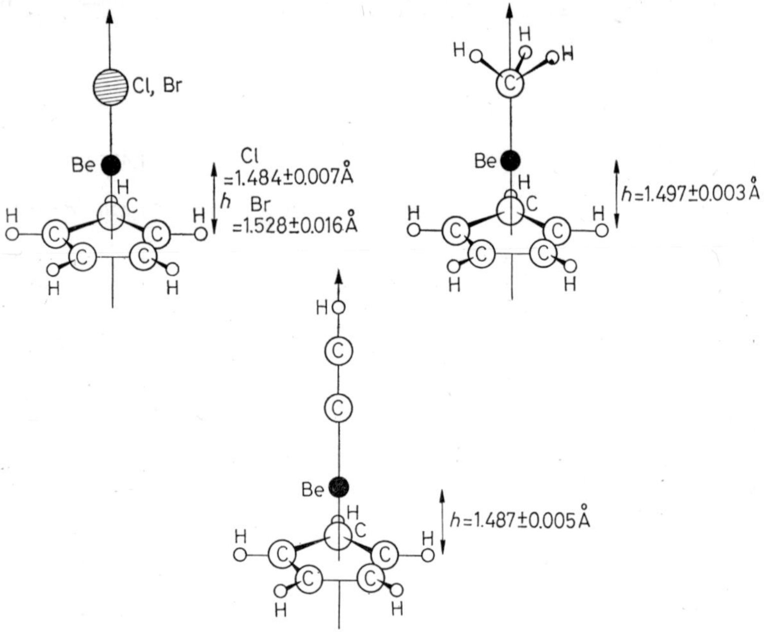

Fig. 65. The molecular models of C_5H_5BeCl (or Br), $C_5H_5BeCH_3$ and $C_5H_5BeC \equiv CH$ (for references see the text)

TABLE 53

Internuclear distance parameters (Å) *for cyclopentadienyl beryllium chloride*

	r_s (a)	r_0 (a)	r_g (b)
$r(C-H)$	1.09 ± 0.03	1.090	1.097 ± 0.004
$r(C-C)$	1.424 ± 0.003	1.424	1.424 ± 0.001
$r(Be-Cl)$	1.81 ± 0.03	1.839	1.837 ± 0.006
$r(Be-ring)$	1.52 ± 0.03	1.485	1.484 ± 0.007
$r(Cl...C)$	3.546 ± 0.005	3.538	3.535 ± 0.005

(a) Microwave spectroscopy, Bjørseth, Drew, Marstokk and Møllendal (1972).
(b) Electron diffraction, Drew and Haaland (1972d).

chlorine and carbon atoms. Thus, the corresponding parameters are particularly accurate and close to the equilibrium values.

The data in Table 52 show that the distance between the beryllium atom and the cyclopentadienyl ring does not vary considerably. In bis(cyclopentadienyl)beryllium, the distance between the beryllium atom and the closest C_5H_5 ring is considered for this comparison. All compounds studied are characterized by approximately the same carbon–carbon intra-ring bond distances. The distance between the beryllium atom and the carbon atom in $CH_3BeC_5H_5$ is the same as the $r(Be-C)$ bond length in dimethylberyllium, $(CH_3)_2Be$. On the other hand, the beryllium–chlorine distance is much larger in C_5H_5BeCl than $r(Be-Cl)$ in $BeCl_2$. A similar though less substantial trend is observed for the values of $r(Be-Br)$ in C_5H_5BeBr and $BeBr_2$. The length of the carbon–carbon triple bond in cyclopentadienylberyllium acetylide ($r_a = 1.231 \pm 0.010$ Å) is not significantly different from that found in acetylene, i.e., $r_g = 1.212 \pm 0.002$ Å (Tanimoto, Kuchitsu and Morino, 1969). Simple hybridization considerations were invoked by Drew and Haaland (1972a, 1972b) in order to explain the geometrical variations in the beryllium derivatives.

BIS(CYCLOPENTADIENYL)MAGNESIUM

Drew and Haaland (1972a) argued that the hybridization model might also be used to explain the asymmetric structure of bis(cyclopentadienyl)beryllium. Accordingly, an asymmetric geometry was expected for the isoelectronic *bis(cyclopentadienyl)magnesium*, $(C_5H_5)_2Mg$. However, as shown by the recent electron diffraction study of Haaland, Lusztyk, Novak, Brunvoll and Starowieyski (1974), the metal atom is equidistant from the two C_5H_5 rings in this compound. In fact, the molecular sandwich type of structure for $(C_5H_5)_2Mg$ had already been established by the early infrared spectroscopic study of a gaseous sample by Cotton and Reynolds (1958).

The distance from the metal atom to the centre of the ring was found to be 2.008 ± 0.004 Å, in terms of r_a parameters. In the course of the electron diffraction structural analysis, the perpendicular amplitude corrections were calculated from a molecular force field, and the refinement of the parameters was carried out for a geometrically consistent r_α structure. The r_a values for the $Mg-C$ and $C-C$ bonds are given in Tables 54 and 55 (see p. 224). These parameters were obtained for the eclipsed model (D_{5h}), which showed better agreement with the experimental data than the staggered model (D_{5d}) did. The large vibrational amplitudes, *viz.*, 0.2 Å, obtained for the distances between different rings indicate that the barrier to internal rotation is less than the thermal energy available (0.8 kcal mol^{-1}), as was emphasized by Haaland, Lusztyk, Novak, Brunvoll and Starowieyski (1974).

DIMETHYL(CYCLOPENTADIENYL)ALUMINIUM

The electron diffraction investigation of *dimethyl(cyclopentadienyl)-aluminium* $(CH_3)_2AlC_5H_5$, by Drew and Haaland (1972b, 1973) showed that the molecule possesses C_s symmetry with a *polyhapto* (*i.e.*, an asymmetrically π-bonded) cyclopentadienyl ring. The models with a *monohapto* (*i.e.*, σ-bonded) ring and a *pentahapto* (*i.e.*, symmetrically π-bonded) ring could be ruled out with confidence. The four possible models with *polyhapto* rings showed equally good agreement with the experimental data. These models are shown in Fig. 66. According to CNDO/2 calculations, model (a) in Fig. 66 is the most likely to be the equilibrium conformation. The geometrical parameters given below were obtained for this model in the electron diffraction structural analysis: the distance between the aluminium atom and the ring = 2.10 $\pm$ 0.02 Å; the distance between the aluminium atom and the symmetry axis of the ring = 0.99 $\pm$ 0.10 Å; $r(C-C)_{cp} = 1.422 \pm 0.002$ Å (the mean intra-ring bond length); and $r(Al-C)_{methyl} = 1.952 \pm 0.003$ Å.

For the symmetrical structure of $CH_3BeC_5H_5$, the cyclopentadienyl ring is considered to be a ligand donating five electrons. Thus, the electron octet around the beryllium atom is provided (Drew and Haaland, 1972c). Similarly, the asymmetric structure of dimethyl-(cyclopentadienyl)aluminium can be interpreted, according to Drew and Haaland (1973), as a consequence of the C_5H_5 ring donating only three electrons to the electron octet of the aluminium atom. No indication could be found, however, of any asymmetry of the ring geometry itself. The $C-C$ bonds seem to be equal (or at least not appreciably different), and the mean value for the $C-C$ bond length is about the same as the value of $r(C-C)$ in $CH_3BeC_5H_5$ (cf., Table 52).

The results of the vapour-phase study on dimethyl(cyclopentadienyl)-aluminium and those of the crystal-phase X-ray diffraction study on

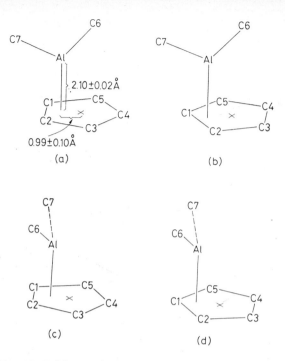

Fig. 66. Polyhapto ring models of dimethyl(cyclopentadienyl)aluminium, $(CH_3)_2AlC_5H_5$ (Drew and Haaland,
1973)

tris(cyclopentadienyl)indium,* $(C_5H_5)_3In$, by Einstein, Gilbert and
Tuck (1972), led Drew and Haaland (1973) to modify the conclusions
drawn earlier for the structure of solid $(CH_3)_2AlC_5H_5$ on the basis
of vibrational spectroscopic data (Haaland and Weidlein, 1972).
According to the latter, there are π-bonded C_5H_5 rings with little
deviation from D_{5h} symmetry in solid $(CH_3)_2AlC_5H_5$ (see Fig. 67a).
It has been recently proposed (Drew and Haaland, 1973) that the
aluminium atoms are connected through bridging C_5H_5 groups, as

* As determined by Einstein, Gilbert and Tuck (1972), tris(cyclopentadienyl)-
indium is polymeric in the solid state. Each indium atom is σ-bonded to two
terminal C_5H_5 groups and is connected with the other indium atoms through
two bridging C_5H_5 groups (cf., Fig. 67b).

217

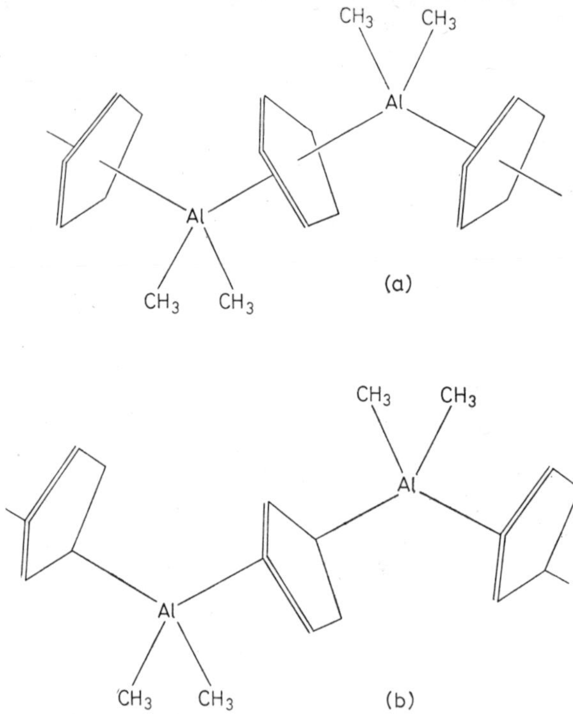

Fig. 67. Proposed molecular structure for solid
dimethyl(cyclopentadienyl)aluminium.
(a) Haaland and Weidlein (1972); (b) Drew and Haaland (1973)

shown in Fig. 67b. Conversely, it has also been suggested (Drew and Haaland, 1973) that the free monomeric molecules of $(C_5H_5)_3In$ would have a structure analogous to that of dimethyl(cyclopentadienyl) aluminium.

INDIUM AND THALLIUM COMPOUNDS

The vapour-phase molecular geometries of *cyclopentadienylindium*, C_5H_5In, and *cyclopentadienylthallium*, C_5H_5Tl, were determined by electron diffraction (Shibata, Bartell and Gavin, 1964) and microwave spectroscopy (Tyler, Cox and Sheridan, 1959), respectively. Both molecules have an "open-faced half-sandwich" configuration with C_{5v}

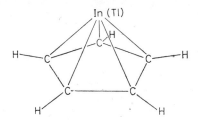

Fig. 68. The "half-sandwich" configuration of (cyclopentadienyl)indium and thallium

symmetry, as illustrated in Fig. 68. The respective values for the carbon–carbon bond distances are 1.426 ± 0.007 Å (r_g) and 1.43 ± 0.02 Å (r_0) for the indium and thallium derivatives. According to the electron diffraction data, the C–H bonds are somewhat bent away from the indium atom out of the plane of the carbon ring. It was the first time that such a phenomenon had been observed for this class of compounds.* The effect (4.5 ± 2°) in C_5H_5In is not necessarily real, but it is in agreement with different considerations on the molecular geometries. Shibata, Bartell and Gavin (1964) give a critical review of the literature on the nature of the bonding in C_5H_5In and C_5H_5Tl and concluded that it is essentially covalent.

TITANIUM AND ZIRCONIUM DERIVATIVES

Bis(cyclopentadienyl)titanium dichloride, $(C_5H_5)_2TiCl_2$ (Alekseev and Ronova, 1966; Ronova and Alekseev, 1967), and *bis(cyclopentadienyl)zirconium dichloride*, $(C_5H_5)_2ZrCl_2$ (Ronova, Alekseev, Gapotchenko and Struchkov, 1970a; 1970b), were studied by electron diffraction. It was found that in both molecules the cyclopentadienyl rings are not parallel. They are eclipsed in the titanium compound and staggered in the zirconium compound, as shown by the molecular models of Fig. 69. Some geometrical parameters are presented in Fig. 69 and in Tables 54 and 55. In addition, the length of the Ti–Cl

---

* The observation was facilitated by the fact that indium is a relatively heavy atom. For example, these features are more difficult to study in the beryllium derivatives as the beryllium atom scatters electrons more weakly.

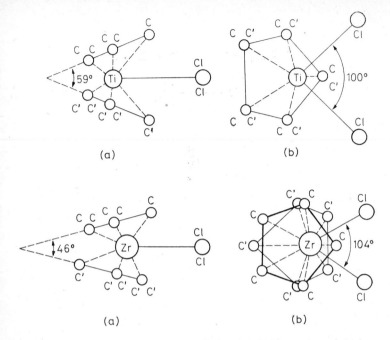

Fig. 69. The molecular configurations of $(C_5H_5)_2TiCl_2$ (Ronova and Alekseev, 1967) and $(C_5H_5)_2ZrCl_2$ (Ronova, Alekseev, Gapotchenko and Struchkov, 1970a). (a) Projection in the plane perpendicular to the C_5H_5 rings; (b) projection in the plane of the chlorine atoms and the metal atom

bond was determined to be 2.24 ± 0.01 Å. Thus it is considerably longer than that in titanium tetrachloride, $r_g(\text{Ti}-\text{Cl}) = 2.170 \pm 0.002$ Å (Morino and Uehara, 1966). The overall molecular geometry of $(C_5H_5)_2TiCl_2$ in the crystalline phase, determined by X-ray diffraction (Tkachev and Atovmyan, 1972), was found to be similar to that of the vapour phase. There are some differences, however, of which the most important can be summarized as follows: (1) the rings were found to be eclipsed in the vapour phase whereas they were staggered in the crystalline phase; (2) the inter-ring angle is 59° in the free molecules and 51° in the crystalline phase; and (3) the Ti−Cl bond was determined to be 0.12 Å longer in the crystalline phase than in the vapour phase. There is no mention in the electron diffraction papers on $(C_5H_5)_2TiCl_2$ of the mean amplitudes of vibration. Both the length

220

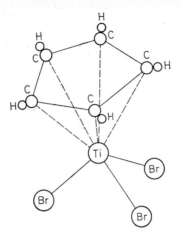

Fig. 70. (Cyclopentadienyl)titanium
tribromide, $C_5H_5TiBr_3$

and the l value were determined for the $Zr-Cl$ bond in $(C_5H_5)_2ZrCl_2$, $r = 2.309 \pm 0.005$ Å and $l = 0.13 \pm 0.005$ Å. The value of $r(Zr-Cl) = 2.32 \pm 0.02$ Å in $ZrCl_4$ (Spiridonov, Akishin and Tsirel'nikov, 1962) is given here for comparison.

The molecular configuration of *cyclopentadienyltitanium tribromide*, $C_5H_5TiBr_3$, was found to be of a "piano-stool shape", as illustrated by Fig. 70. The geometrical parameters determined by the electron diffraction study of Ronova and Alekseev (1969) include $r(Ti-C)$, given in Table 54, and $r(Ti-Br) = 2.310 \pm 0.005$ Å. In $TiBr_4$, the $Ti-Br$ distance was given by Lister and Sutton (1941) as $r(Ti-Br) = 2.31 \pm 0.02$ Å. The $Br-Ti-Br$ bond angle was found to be between 100.2° and 102.3°. The results of the vapour-phase investigation are in agreement with the findings of crystal-phase X-ray diffraction work on $C_5H_5TiCl_3$ by Ganis and Allegra (1963).

LEAD AND TIN COMPOUNDS

Some of the geometrical parameters of *bis(cyclopentadienyl)lead*, $(C_5H_5)_2Pb$, and *bis(cyclopentadienyl)tin*, $(C_5H_5)_2Sn$, are presented in Tables 54 and 55. The structural determinations were performed by electron diffraction (Almenningen, Haaland and Motzfeldt, 1967a). In both instances the ligand rings are not parallel, the angles between their planes being $45 \pm 15°$ and $55°$ in $(C_5H_5)_2Pb$ and $(C_5H_5)_2Sn$, respectively. The relative orientations of the rings could not be determined, but it was assumed that there is free or only slightly hindered rotation. The results obtained for the tin compound are more uncertain than those for the lead derivative because of experimental difficulties.

CYCLOPENTADIENYL AND BENZENE COMPLEXES OF CHROMIUM

An electron diffraction investigation of *bis(cyclopentadienyl)chromium*, $(C_5H_5)_2Cr$, was carried out by Haaland, Lusztyk, Novak, Brunvoll and Starowieyski (1974) under conditions essentially identical with those for bis(cyclopentadienyl)magnesium. The distance from the chromium atom to the centre of the ring was found to be 1.798 ± 0.004 Å. The r_a values for the $Cr-C$ and $C-C$ bonds obtained for the eclipsed model are given in Tables 54 and 55. While no significant deviation from planarity was found for the C_5H_5 rings in $(C_5H_5)_2Mg$, the $C-H$ bonds are bent $2.9 \pm 1.1°$ out of the plane of the carbon skeleton towards the metal atom in $(C_5H_5)_2Cr$.

As *dibenzene chromium*, $(C_6H_6)_2Cr$, is the only aryl derivative whose molecular geometry has been determined in the gaseous phase, it was considered appropriate to discuss it in this section. The main purpose of the electron diffraction investigation by Haaland (1965) was to determine whether the C_6H_6 ligand in the complex retains the D_{6h} symmetry of the free benzene molecule. Contradictory results concerning this question had appeared earlier, originating from both spectroscopic (Fritz and Lüttke, 1959; Fritz, Lüttke, Stammreich and Forneris, 1959) and X-ray diffraction measurements (Jellinek, 1960, 1963; Cotton, Dollase and Wood, 1963) referring to the solid-state structure. The electron diffraction radial distribution could be interpreted by a model with D_{6h} symmetry such as that shown in Fig.71. It was found that the mean amplitude of vibration for the carbon–carbon bonds in $(C_6H_6)_2Cr$ is the same as that in free benzene. This is important evidence for the concept that there is only one type of carbon–carbon bond in the complex molecule. More recent data obtained by other physical techniques are in complete agreement with the

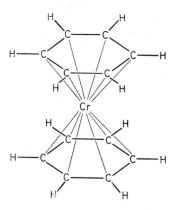

Fig. 71. The molecular model with D_{6h} symmetry for $(C_6H_6)_2Cr$

findings of the electron diffraction study concerning the D_{6h} symmetry of the molecule. These studies include a crystal structure determination by Keulen and Jellinek (1966) and an infrared spectroscopic investigation by Ngai, Stafford and Schäfer (1969). Further evidence was produced by a thermodynamic study (Andrews, Westrum and Bjerrum, 1969) and a normal coordinate analysis (Schäfer, Southern, Cyvin and Brunvoll, 1970). The latter utilized both the infrared (Ngai, Stafford and Schäfer, 1969) and more recent Raman (Schäfer, Southern and Cyvin, 1971) measurements. The calculated mean amplitudes of vibration for the model of $(C_6H_6)_2Cr$ with D_{6h} symmetry were in excellent agreement with the electron diffraction results. Some more complete calculations have also appeared since then, showing full agreement with the earlier results (Brunvoll, Cyvin and Schäfer, 1971; Cyvin, Brunvoll and Schäfer, 1971).

The length of the carbon–carbon bond itself in $(C_6H_6)_2Cr$, $r_g =$ = 1.424 ± 0.002 Å, is larger than that in free benzene ($r_g = 1.399$ Å) determined by Almenningen, Bastiansen and Fernholt (1958). The value of $r_g(Cr-C)$ was found to be 2.152 ± 0.002 Å.

TABLE 54

*Parameters for the metal–carbon bonds (Å) in cyclopentadienyl complexes**

Compounds	r(M—C)	l(M—C)	$r_{cov.}$ (a)	$\Delta_{cov.}$ (b)	r_{ion} (a)	Δ_{ion} (b)
Be(C$_5$H$_5$)$_2$	1.907±0.007	0.098±0.004	0.889	1.018	0.34	1.57
	2.256±0.007	0.115±0.007				
Mg(C$_5$H$_5$)$_2$	2.339±0.004		1.364	0.982	0.78	1.57
In(C$_5$H$_5$)	2.621±0.005	0.077±0.007	1.497	1.124	1.32	1.30
Tl(C$_5$H$_5$)	2.705±0.005		1.549	1.156	1.40	1.31
Ti(C$_5$H$_5$)$_2$Cl$_2$	2.38 ±0.01		1.324	1.06	0.90	1.48
Ti(C$_5$H$_5$)Br$_3$	2.336±0.01		1.324	1.012	0.90	1.44
Zr(C$_5$H$_5$)$_2$Cl$_2$	2.522±0.005	0.11 ±0.005	1.454	1.068		
Sn(C$_5$H$_5$)$_2$	2.706±0.008	0.156±0.027	1.399	1.307	1.12	1.59
Pb(C$_5$H$_5$)$_2$	2.778±0.005	0.142±0.013	1.538	1.240	1.20	1.58
Cr(C$_5$H$_5$)$_2$	2.169±0.004		1.176	0.989	0.84	1.33
Mn(C$_5$H$_5$)$_2$	2.383±0.003	0.135±0.005	1.171	1.212	0.80	1.58
Fe(C$_5$H$_5$)$_2$	2.063±0.003	0.062±0.001	1.165	0.898	0.76	1.30
Ru(C$_5$H$_5$)$_2$	2.196±0.003	0.060±0.001	1.246	0.950		
Ni(C$_5$H$_5$)$_2$	2.196±0.008	0.084±0.006	1.154	1.042	0.72	1.48

* For references, see text.

(a) $r_{cov.}$ = covalent radii and r_{ion} = radii of the dipositive (for In and Tl, monopositive) metal ions from Pauling (1960).

(b) $\Delta_{cov.} = r(M—C) - r_{cov.}$; $\Delta_{ion} = r(M—C) - r_{ion}$.

TABLE 55

*Lengths (r) and mean amplitudes of vibration (l) for the carbon–carbon bonds in cyclopentadienyl complexes**

	r(C—C), Å	l(C—C), Å
Be(C$_5$H$_5$)$_2$	1.425±0.002	0.052±0.001
Mg(C$_5$H$_5$)$_2$	1.424±0.002	
Al(C$_5$H$_5$)(CH$_3$)$_2$	1.422±0.002	0.049±0.002
In(C$_5$H$_5$)	1.427±0.007	0.040±0.009
Tl(C$_5$H$_5$)	1.43 ±0.02	
Ti(C$_5$H$_5$)$_2$Cl$_2$	1.40	
Ti(C$_5$H$_5$)Br$_3$	1.41 ±0.02	
Zr(C$_5$H$_5$)$_2$Cl$_2$	1.42 ±0.01	0.05 ±0.005
Sn(C$_5$H$_5$)$_2$	1.431±0.009	0.044±0.019
Pb(C$_5$H$_5$)$_2$	1.430±0.007	0.045±0.005
Cr(C$_5$H$_5$)$_2$	1.432±0.002	
Mn(C$_5$H$_5$)$_2$	1.429±0.005	0.052±0.002
Fe(C$_5$H$_5$)$_2$	1.440±0.002	0.046±0.001
Ru(C$_5$H$_5$)$_2$	1.439±0.002	0.051±0.001
Ni(C$_5$H$_5$)$_2$	1.430±0.003	0.044±0.003

* For references, see text.

BIS(CYCLOPENTADIENYL)MANGANESE

Free rotation of the cyclopentadienyl rings with a parallel relative orientation was found for bis(cyclopentadienyl)manganese, $(C_5H_5)_2Mn$, in an electron diffraction study by Almenningen, Haaland and Motzfeldt (1967b). The effect of assuming harmonic vibrations on the determination of the geometrical parameters was also examined. Some of the structural parameters determined are presented in Tables 54 and 55.

FERROCENE

Several electron diffraction structural determinations have been performed for *bis(cyclopentadienyl)iron* (or *ferrocene*), $(C_5H_5)_2Fe$. The most complete work, by Haaland and Nilsson (1968b), was partly based on the experimental data obtained by Bohn and Haaland (1966), although some that were found to suffer from a slight experimental scale error were eliminated. The earlier studies were carried out by Seibold and Sutton (1955) and Akishin, Rambidi and Bredikhina (1961). In addition to detailed studies on the molecular geometry, the internal motion and the barrier to internal rotation were examined extensively by Haaland and Nilsson (1968a, 1968b).

Of the compounds dealt with in this chapter, bis(cyclopentadienyl)-iron was the first member to be synthesized (Miller, Tebboth and Tremaine, 1952; Kealy and Pauson, 1951). Its novel molecular configuration was soon established accurately by Fischer and Phab (1952) and Wilkinson, Rosenblum, Whiting and Woodward (1952), primarily on the basis of spectroscopic data. The five-fold symmetry axes of the two regular and parallel C_5H_5 rings coincide. As far as the relative orientation of the rings is concerned, the molecule may be *prismatic* (eclipsed conformation) or *antiprismatic* (staggered conformation), as demonstrated in Fig. 72. The corresponding symmetries are D_{5h} and D_{5d} for the *prismatic* and *antiprismatic* models, respectively. It was not possible to decide between the two forms on the basis of only spectroscopic information (Lippincott and Nelson, 1958; Fritz, 1964). As the ferrocene crystals are highly disordered (cf., Haaland and Nilsson, 1968b), no conclusion on the molecular symmetry could be drawn from the crystalline structure determinations. The two early electron diffraction studies (Seibold and Sutton, 1955; Akishin,

15

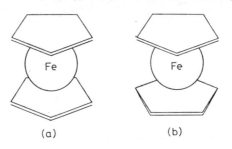

Fig. 72. (a) *Prismatic* and (b) *antiprismatic* models of $(C_5H_5)_2Fe$

Rambidi and Bredikhina, 1961) suggested the existence of free rotation of the cyclopentadienyl rings. On the other hand, Bohn and Haaland (1966) found that the *prismatic* form was more stable, with a potential barrier of 1.1 kcal mol^{-1}.

In order to determine the potential barrier more accurately, the molecular intensity of electron scattering was given by Haaland and Nilsson (1968a, 1968b) as a function of the angular probability distribution function $W(\varphi)$. The latter is defined in such a way that $W(\varphi)d\varphi$ is the probability of finding the angle of rotation φ in the interval between φ and $\varphi + d\varphi$. The classical expression $W(\varphi) = N\exp\left(-\dfrac{V(\varphi)}{kT}\right)$ was proved to be satisfactory for use in the calculations. Here the potential energy is given as $V(\varphi) = (V_0/2)(1 - \cos 5\varphi)$ and V_0 is the barrier height. The barrier to internal rotation in ferrocene was found to be $V_0 = 0.9 \pm 0.3$ kcal mol^{-1} as a result of the least-squares refinements based on the molecular intensities, with the prismatic form being more stable. A typical curve for the $W(\varphi)$ function obtained as a result of the calculations is shown in Fig. 73. Some of the structural parameters determined are given in Tables 54 and 55.

Bohn and Haaland (1966) determined the deviation from planarity for the C_5H_5 ring in $(C_5H_5)_2Fe$. According to the Fe ... H distance, the angle by which the C$-$H bonds are bent out of the plane of the carbon ring skeleton towards the iron atom was found to be $4.6 \pm 0.9°$. As the perpendicular amplitude corrections have recently been calculated for $(C_5H_5)_2Fe$ by Schäfer, Brunvoll and Cyvin (1972), the

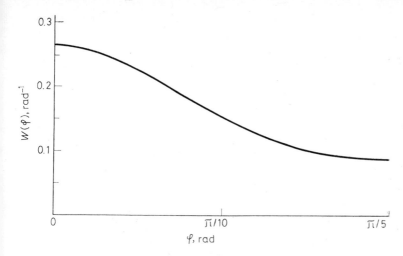

Fig. 73. Typical angular probability distribution function $W(\varphi)$ as plotted against the angle of rotation φ around the five-fold symmetry axis of ferrocene (Haaland and Nilsson, 1968b)

$\varphi = 0$, *prismatic* and $\varphi = \dfrac{\pi}{5}$ rad, *antiprismatic* conformations

calculated shrinkage for the Fe . . . H distance was 0.010 Å. The corresponding angle of bending is $3.7 \pm 0.9°$, as given by Haaland, Lusztyk, Novak, Brunvoll and Starowieyski (1974).

RUTHENIUM AND NICKEL COMPOUNDS

Bis(cyclopentadienyl)ruthenium (or *ruthenocene*), $(C_5H_5)_2Ru$, was studied simultaneously with ferrocene by Haaland and Nilsson (1968b). The experimental electron diffraction data were best reproduced by a *prismatic* model. Attempts to determine the barrier to internal rotation were unsuccessful, which was attributed to the experimental data being less accurate than those for ferrocene. Some of the structural parameters determined are given in Tables 54 and 55.

Ronova, Bochvar, Chistjakov, Struchkov and Alekseev (1969) and also Hedberg and Hedberg (1970) applied electron diffraction to determine the molecular structure of *bis(cyclopentadienyl)nickel* (or *nickelocene*), $(C_5H_5)_2Ni$. The results of the two studies show disagreement firstly concerning the size of the molecule. Hedberg and Hedberg (1970) showed that their data are to be preferred. In addition to the disagreement on the molecular size, the following important differences should also be mentioned. According to Ronova, Bochvar, Chistjakov, Struchkov and Alekseev (1969), the molecular conformation of $(C_5H_5)_2Ni$ is eclipsed and the C−H bonds are bent from the plane of the carbon atom ring skeleton towards the metal atom. On the other hand, the results of Hedberg and Hedberg (1970) show that the C_5H_5 rings are freely rotating and the deviation of the C−H bonds from the C_5 plane cannot be considered significant. There is no mention of the vibrational parameters in the former work although they are important in the structural analysis and discussed in detail in the latter paper. The most important structural data according to Hedberg and Hedberg (1970) are given in Tables 54 and 55.

The bond distances of cyclopentadienyl nitrosyl nickel, C_5H_5NiNO,

as determined by microwave spectroscopy (Cox and Brittain, 1970) are as follows:

$$r(C-C) = 1.43 \ \text{Å}$$
$$r(Ni=N) = 1.626 \ \text{Å}$$
$$r(N=O) = 1.165 \ \text{Å}.$$

COMPARISON OF METALLOCENE STRUCTURES

The C_5H_5 rings possess C_5 symmetry in the cyclopentadienyl complexes. Even in dimethyl(cyclopentadienyl)aluminium with *polyhapto* bonded rings, the mean value for the carbon–carbon bond distances is the same as in the other complexes. This indicates a symmetrical or only slightly distorted ring configuration. There is little variation of $r(C-C)$ in the different compounds, as can be concluded from comparing the data in Table 55. For instance, the carbon–carbon bond is longer by 0.01 Å in the more strongly bonded ferrocene than in the more weakly bonded nickelocene. Similarly, the $C-C$ bond in dibenzene chromium is longer than that in free benzene.

The metal–carbon bond lengths are the most important parameters in characterizing the strength of the coordination bonding. These values, together with the corresponding mean amplitudes of vibration, are given in Table 54. The covalent radii and the radii of the dipositive (or monopositive) metal ions (Pauling, 1960) are also given.

The short metal–carbon bonds appear, *e.g.*, in ferrocene, together with small amplitudes of vibration, indicating a strong linkage between the cyclopentadienyl rings and the metal atom. Although the covalent radii of nickel and iron are essentially the same (cf., Table 54), the $Ni-C$ bond in nickelocene is 0.14 Å longer than the $Fe-C$ bond in ferrocene. Also, the mean amplitude of vibration of the nickel–carbon bond is considerably larger than the value of $l(Fe-C)$. In addition, the $C-C$ bond is shorter in $(C_5H_5)_2Ni$ than that in $(C_5H_5)_2Fe$, and the bending of the $C-H$ bonds from the C_5 plane is smaller in nickelocene. It can be concluded that the intra-ring bonding is stronger and the linkage between the metal atom and the rings is weaker in the nickel derivative. This conclusion is also supported by the large difference in the l values in $(C_5H_5)_2Ni$ and $(C_5H_5)_2Fe$ for non-bond distances. The respective l values for nickelocene are the larger. Such a trend is shown by the data both as determined from the electron diffraction experiment and as calculated from the vibrational spectroscopic analysis given in Table 56. The numbering of the atoms is presented in Fig. 74.

The difference between the molecular geometries of $(C_5H_5)_2Fe$ and $(C_5H_5)_2Mn$ follows the same trend, but it is even stronger than that described above for ferrocene and nickelocene. The covalent radii of iron and manganese are again the same, but the difference in the lengths of the metal–carbon bonds is 0.32 Å. Further, the mean amplitude of vibration for the $Mn-C$ bond is more than twice the value of $l(Fe-C)$. The rather loose coordination linkage in the manganese derivative is also well characterized by the large l values for atomic pairs in which the atoms belong to different cyclopentadienyl rings (Almenningen, Haaland and Motzfeldt, 1967b):

$$l(\text{C1} \ldots \text{C6}) = 0.259 \pm 0.112 \text{ Å}$$
$$l(\text{C1} \ldots \text{C7}) = 0.356 \pm 0.066 \text{ Å}$$
$$l(\text{C1} \ldots \text{C8}) = 0.231 \pm 0.046 \text{ Å}.$$

The numbering of atoms is given in Fig. 74. The coordination linkage is similarly loose in the cyclopentadienyl derivatives of lead and tin, as demonstrated by the large differences between the bond distance $r(\text{M}-\text{C})$ and the covalent radius in both instances and by the large values of $l(\text{M}-\text{C})$.

Haaland (1969) classified $(C_5H_5)_2Fe$, $(C_5H_5)_2Ru$ and $(C_6H_6)_2Cr$ as "strongly bonded complexes" and $(C_5H_5)_2Mn$, $(C_5H_5)_2Pb$, $(C_5H_5)_2Sn$ and $(C_5H_5)_2Be$ as "weakly bonded complexes". He made this classification strictly on the basis of the geometrical data, emphasizing that it did not necessarily imply information on the nature of bonding in this series of compounds. As new experimental data are now available for more compounds, these two groups of compounds can be augmented. $(C_5H_5)_2Mg$ and $(C_5H_5)_2Cr$ are referred to as "strongly bonded

TABLE 56

Mean amplitudes of vibration for $(C_5H_5)_2Fe$ and $(C_5H_5)_2Ni$ as determined by electron diffraction (ED) and as calculated from spectroscopic data (SP)

Atomic pairs*	$(C_5H_5)_2Fe$			$(C_5H_5)_2Ni$		
	r, Å	$l(SP)$, Å	$l(ED)$, Å	r, Å	$l(SP)$, Å	$l(ED)$, Å
		(a)	(b)		(c)	(d)
M−C	2.06	0.083	0.062 ± 0.001	2.20	0.110	0.084 ± 0.006
C−C	1.43	0.047	0.046 ± 0.001	1.43	0.047	0.044 ± 0.003
C−H	1.12	0.077	0.092 ± 0.006	1.08	0.079	0.079 ± 0.023
M...H	2.86	0.124	0.124 ± 0.009	2.94	0.146	0.147 ± 0.035
C1...C3	2.31	0.064	0.054 ± 0.002	2.31	0.064	0.054 ± 0.006
C1...H2	2.27	0.100	0.145 ± 0.015	2.24	0.100	(0.090)
C1...H3	3.39	0.098	0.158 ± 0.017	3.36	0.098	(0.100)
C1...C6	3.32	0.159	0.101 ± 0.008		0.212	(0.200)
C1...C7	3.61	0.135	0.193 ± 0.019		0.177	(0.180)
C1...C8	4.04	0.092	0.117 ± 0.020		0.119	(0.160)

* The numbering of atoms is given in Fig. 74.
(a) Schäfer, Brunvoll and Cyvin (1972).
(b) Haaland and Nilsson (1968b).
(c) Brunvoll, Cyvin, Ewbank and Schäfer (1972).
(d) Hedberg and Hedberg (1970).

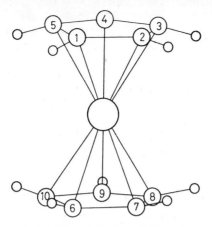

Fig. 74. The numbering of atoms in the
$(C_5H_5)_2M$ sandwich molecule

complexes" while C_5H_5BeX (X = CH_3, C≡CH, Cl, Br or BH_4), C_5H_5In, C_5H_5Tl, titanium and zirconium derivatives and $(C_5H_5)_2Ni$ are "weakly bonded complexes". In the extreme cases, the correlation is obvious between the geometrical variations and bonding peculiarities deduced from various physico-chemical data (Fischer and Fritz, 1959; Wilkinson and Cotton, 1959). Ferrocene and ruthenocene are considered to be covalent complexes. An important role is attributed to the *d* orbitals of the metal atom in forming the π-bonding in $(C_5H_5)_2Fe$. The *d* orbitals in lead, tin or beryllium derivatives are not considered to be available. These compounds are regarded as intermediates between the ionic and covalent types and were named "centrally σ-bonded" (Fritz, 1964). From the overall picture, it can certainly be concluded that the information on the geometrical variations often provides an insight into the nature of bonding, as was seen for ferrocene, for example.

The presence of a weaker coordination linkage in $(C_5H_5)_2Ni$ probably does not originate from a fundamental change in the type of bonding as in $(C_5H_5)_2Fe$. As was suggested by Hedberg and Hedberg (1970), in nickelocene the pair of electrons, additional to the 18 in $(C_5H_5)_2Fe$ (five from each of the rings and eight from the iron atom), goes into antibonding orbitals and consequently the bonding becomes weaker.

The comparisons involving the ionic radius and its difference from the $r(M-C)$ bond length (cf., Table 54) may also be helpful in seeking correlations between bonding and the molecular geometry. The constancy of the difference $r(M-C) - r_{ion}$, for example, seems to be a typical feature in the series $(C_5H_5)_2Be$, $(C_5H_5)_2Mg$, $(C_5H_5)_2Mn$, $(C_5H_5)_2Sn$ and $(C_5H_5)_2Pb$. Here the shorter of the two types of beryllium–carbon bonds in beryllocene was taken into consideration. Haaland (1969) drew attention to the fact that if both rings were this close ($h = 1.47$ Å) to the beryllium atom in $(C_5H_5)_2Be$, it would result in a ring-to-ring distance of 2.94 Å. This distance is much smaller than the van der Waals "radius" (or rather, van der Waals thickness) of the aromatic systems (3.4 Å, Pauling, 1960). The actual geometry, characterized by a ring-to-ring distance of 3.375 ± 0.010 Å, indicates that the van der Waals forces are the determining factor. The binding energy between the cyclopentadienyl ring and the beryllium atom has its minimum at a beryllium–ring distance of 1.47 Å, as suggested not only by the shorter $Be-C$ bond in $(C_5H_5)_2Be$ but also by the values of $r(Be-C)$ in the half-sandwich type beryllium derivatives (see Table 52). The sum of the van der Waals "radii" of the two cyclopentadienyl rings is considerably larger than twice the distance between the beryllium atom and the C_5H_5 ring corresponding to the minimum energy. The repulsive interactions between the π-electron systems of the two rings are decisive in establishing the size of the bis(cyclopentadienyl)-beryllium molecule. The beryllium atom is not a determining factor because of its small size. To a certain extent, this small atom moves unrestricted in the cavity formed between the two rings. On the other hand, with a large metal atom the molecular geometry, *i.e.*, the ring-to-ring distance, is determined by the ionic radius of the metal atom. The latter is the case for bis(cyclopentadienyl)magnesium and for all other molecules where the difference $r(M-C) - r_{ion}$ is 1.57–1.58 Å (cf., Table 54). Molecular orbital calculations utilizing various assumptions (Sundbom, 1966; Lopatko, Klimenko and Dyatkina, 1972) provide additional evidence for the asymmetric molecular geometry being energetically more favourable than the symmetrical configuration.

It is of interest to compare the data on the relative orientations of the two cyclopentadienyl (or benzene) rings in symmetrically π-bonded complexes. The rings are parallel in most instances. The exceptions are titanium and zirconium derivatives that also contain other ligands,

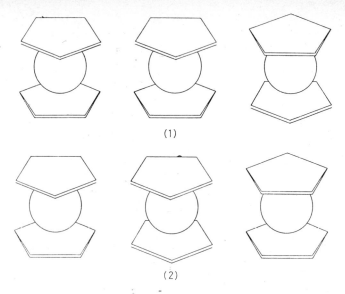

(1)

(2)

Fig. 75. Two model mixtures for ferrocene, consistent with the
X-ray diffraction data (after Haaland and Nilsson, 1968b)

and lead and tin compounds. In all complexes, including those with
non-parallel rings, the metal atoms are situated on the symmetry
axes of the rings.

Eclipsed *(prismatic)* conformations of the two cyclopentadienyl
rings were found in the "strongly bonded complexes" (Haaland,
1969). This finding agrees well with the results of a crystal-phase
X-ray diffraction study on ruthenocene (Hardgrave and Templeton,
1959). The data referring to the gaseous (Haaland, 1965) and crystalline (Jellinek, 1960; 1963; Cotton, Dollase and Wood, 1963; Keulen
and Jellinek, 1966) molecular structures are also in complete agreement concerning the *prismatic* conformation of dibenzene chromium.
As has already been mentioned, the crystals of ferrocene are disordered.
The X-ray and neutron diffraction data are compatible with both combinations of models presented in Fig. 75. The intermolecular forces
are probably of great importance in forming these conformations.
The forms observed for crystalline diferrocyl (Kaluski, Struchkov
and Avoyan, 1964; Macdonald and Trotter, 1964) and terferrocyl
(Kaluski and Struchkov, 1965) may provide some insight into these

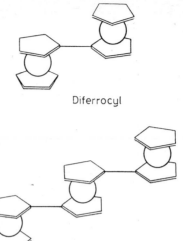

Diferrocyl

Terferrocyl

Fig. 76. The crystal-phase molecular
configuration of (a) diferrocyl and (b)
terferrocyl (for references see the text)

conformational problems (Fig. 76). Attention was drawn to them
by Haaland and Nilsson (1968b) in this connection. The ferrocyl units
are of intermediate structure between the *prismatic* and *antiprismatic*
forms, whereas in terferrocyl the first and third units are *prismatic*
with an *antiprismatic* unit between them. An *antiprismatic* conformation has been attributed to the ferrocene molecules for a long time.
Finally, it was Bohn and Haaland (1966) and Haaland and Nilsson
(1968b) who firmly established the *prismatic* conformation of the free
$(C_5H_5)_2Fe$ molecules. It is appropriate to mention that Wilkinson,
Rosenblum, Whiting and Woodward (1952), when describing the
structure of ferrocene for the first time, stressed that the *prismatic*
structure could not be excluded, although the experimental evidence
available at that time provided strong support for the *antiprismatic*
form.

The two rings rotate freely in bis(cyclopentadienyl)manganese
according to the electron diffraction data (Almenningen, Haaland
and Motzfeldt, 1967b). An *antiprismatic* conformation was observed
for bis(cyclopentadienyl)beryllium (Almenningen, Bastiansen and

Haaland, 1964; Haaland, 1968). The two rings are 4.10 Å apart in $(C_5H_5)_2Mn$, and thus the van der Waals repulsion between the rings is negligible. On the other hand, the interaction between the rings may well be the determining factor in the conformational choice for $(C_5H_5)_2Be$. It was suggested by Haaland (1969) that a strong correlation exists between the occurrence of the prismatic conformation and the ligand–metal–ligand bond system in the "strongly bonded complexes".

As the comparisons between the molecular geometries of the free and complexed molecules provide useful information in studying the nature of bonding in the coordination compounds, it may be of similar value to perform such comparisons concerning the intramolecular motion. Some of the first subjects in the investigations on the influence of complexation on the vibrational behaviour of molecules were performed on compounds belonging to the class discussed here. These include ferrocene (Schäfer Brunvoll and Cyvin, 1972), dibenzene chromium (Cyvin, Brunvoll and Schäfer, 1971), benzene tricarbonyl chromium (Brunvoll, Cyvin and Schäfer, 1972) and analogous systems (Cyvin, Cyvin and Brunvoll, 1970; Schäfer, Ewbank, Cyvin and Brunvoll, 1972).

REFERENCES*

Abrahams, S. C. and Bernstein, J. L. (1966): J. Chem. Phys. **45**, 2745.
Abrahams, S. C. and Reddy, J. M. (1965): J. Chem. Phys. **43**, 2533.
Ackerman, R. J. and Rauh, E. G. (1963): J. Phys. Chem. **67**, 2596.
Adams, D. M. and Squire, A. (1968): J. Chem. Soc. A, 2817.
Adams, W. J. and Bartell, L. S. (1971): J. Mol. Struct. **8**, 23.
Addison, C. C. and Hathaway, B. J. (1958): J. Chem. Soc., 3099.
Adrian, H. W. W. and Feil, D. (1969): Acta Cryst. **A25**, 438.
Afonskii, N. S. (1962): Zh. Neorg. Khim. **7**, 2640.
Ahlrichs, R. (1973): Chem. Phys. Letters **19**, 174.
Akischin, P. A. and Rambidi, N. G. (1960): Z. phys. Chem. **213**, 111.
Akishin, P. A., Naumov, V. A. and Tatevskii, V. M. (1959): Kristallografija **4**, 194.
Akishin, P. A. and Rambidi, N. G. (1958a): Vest. Moskov. Univ. Ser. Khim. No. 6, 223.
Akishin, P. A. and Rambidi, N. G. (1958b): Zh. Neorg. Khim. **3**, 2599.
Akishin, P. A. and Rambidi, N. G. (1959): Zh. Neorg. Khim. **4**, 718.
Akishin, P. A. and Rambidi, N. G. (1960): Zh. Neorg. Khim. **5**, 23.
Akishin, P. A., Rambidi, N. G. and Bredikhina, T. N. (1961): Zh. Strukt. Khim. **2**, 476.
Akishin, P. A., Rambidi, N. G., Kuznetsov, G. N. and Matrosov, E. I. (1957): Zh. Neorg. Khim. **2**, 1699.
Akishin, P. A., Rambidi, N. G. and Spiridonov, V. P. (1967): High-Temperature Electron Diffraction by Gases. The Characterization of High-Temperature Vapors (ed.: Margrave, J. L.). Wiley, New York.
Akishin, P. A., Rambidi, N. G. and Zasorin, E. Z. (1959a): Kristallografija **4**, 186.
Akishin, P. A., Rambidi, N. G. and Zasorin, E. Z. (1959b): Kristallografija **4**, 360.
Akishin, P. A. and Spiridonov, V. P. (1957): Kristallografija **2**, 475.
Akishin, P. A. and Spiridonov, V. P. (1960): Dokl. Akad. Nauk S. S. S. R. **131**, 557.
Akishin, P. A. and Spiridonov, V. P. (1961a): Zh. Strukt. Khim. **2**, 63.
Akishin, P. A. and Spiridonov, V. P. (1961b): Zh. Strukt. Khim. **2**, 542.

* The spelling of the names of Soviet authors may be different in various places. The names from the Russian language references were transliterated according to the British Standard. However, the spelling of the names from papers published, *e.g.*, in German or Hungarian, was not changed so as to facilitate literature searches. Cf., *e.g.*, Akishin and Akischin; Spiridonov and Szpiridonov.

Akishin, P. A. and Spiridonov, V. P. (1962): Zh. Strukt. Khim. **3**, 267.

Akishin, P. A., Vilkov, L. V. and Rosolovskii, V. Ya. (1959): Kristallografija **4**, 353.

Akishin, P. A., Vilkov, L. V. and Rosolovskii, V. Ya. (1960): Zh. Strukt. Khim. **1**, 5.

Akishin, P. A., Vinogradov, M. I., Danilov, K. D., Levkin, N. P., Martinson, E. N., Rambidi, N. G. and Spiridonov, V. P. (1958): Prib. Tekhn. Eksp. 70.

Alekseev, N. V. and Ronova, I. A. (1966): Zh. Strukt. Khim. **7**, 103.

Alfheim, I., Hagen, G. and Cyvin, S. J. (1971): J. Mol. Struct. **8**, 159.

Allegra, G., Perego, G. and Immirzi, A. (1963): Macromol. Chem. **61**, 69.

Almenningen, A., Andersen, B. and Astrup, E. E. (1970): Acta Chem. Scand. **24**, 1579.

Almenningen, A., Anderson, G. A., Forgaard, F. R. and Haaland, A. (1971): Chem. Commun., 480.

Almenningen, A., Anderson, G. A., Forgaard, F. R. and Haaland, A. (1972): Acta Chem. Scand. **26**, 2315.

Almenningen, A., Bastiansen, O. and Fernholt, L. (1958): Kgl. Norske Videnskab. Selskabs. Skrifter No. 3.

Almenningen, A., Bastiansen, O. and Haaland, A. (1964): J. Chem. Phys. **40**, 3434.

Almenningen, A., Bastiansen, O. and Motzfeldt, T. (1969): Acta Chem. Scand. **23**, 2848.

Almenningen, A., Bastiansen, O. and Motzfeldt, T. (1970): Acta Chem. Scand. **24**, 747.

Almenningen, A., Gundersen, G. and Haaland, A. (1967): Chem. Commun., 557.

Almenningen, A., Gundersen, G. and Haaland, A. (1968a): Acta Chem. Scand. **22**, 328.

Almenningen, A., Gundersen, G. and Haaland, A. (1968b): Acta Chem. Scand. **22**, 859.

Almenningen, A., Gundersen, G., Haugen, T. and Haaland, A. (1972): Acta Chem. Scand. **26**, 3928.

Almenningen, A., Haaland, A., Haugen, T. and Novak, D. P. (1973): Acta Chem. Scand. **27**, 1821.

Almenningen, A., Haaland, A. and Morgan, G. L. (1969): Acta Chem. Scand. **23**, 2921.

Almenningen, A., Haaland, A. and Motzfeldt, T. (1967a): J. Organometal. Chem. **7**, 97.

Almenningen, A., Haaland, A. and Motzfeldt, T. (1967b): On the Molecular Structure of Manganese Dicyclopentadienide, $Mn(C_5H_5)_2$. Selected Topics in Structure Chemistry (ed.: Andersen, P., Bastiansen, O. and Furberg, S.). Universitetsforlaget, Oslo.

Almenningen, A., Haaland, A. and Wahl, K. (1969a): Acta Chem. Scand. **23**, 1145.

Almenningen, A., Haaland, A. and Wahl, K. (1969b): Acta Chem. Scand. **23**, 2245.

Almenningen, A., Halvorsen, S. and Haaland, A. (1971): Acta Chem. Scand. **25**, 1937.

238

Almenningen, A., Jacobsen, G. G. and Seip, H. M. (1969): Acta Chem. Scand. 23, 685.

Amano, T., Hirota, E. and Morino, Y. (1967): J. Phys. Soc. Japan 22, 399.

Anderson, G. A., Forgaard, F. R. and Haaland, A. (1972): Acta Chem. Scand. 26, 1947.

Andrews, J. T. S., Westrum, E. F. and Bjerrum, N. (1969): J. Organometal. Chem. 17, 293.

Andrianov, V. G., Biryukov, B. P. and Struchkov, Yu. T. (1969): Zh. Strukt. Khim. 10, 1129.

Arnesen, S. P. and Seip, H. M. (1966): Acta Chem. Scand. 20, 2711.

Atoji, M. and Lipscomb, W. N. (1954): Acta Cryst. 7, 173.

Atoji, M., Wheatley, P. J. and Lipscomb, W. N. (1955): J. Chem. Phys. 23, 1176.

Atoji, M., Wheatley, P. J. and Lipscomb, W. N. (1957): J. Chem. Phys. 27, 196.

Ault, B. S. and Pimentel, G. C. (1973): J. Phys. Chem. 77, 1649.

Bach, M.-C., Crasnier, F., Labarre, J.-F. and Leibovici, C. (1973): J. Mol. Struct. 16, 89.

Badger, R. M. (1934): J. Chem. Phys. 2, 128.

Bailey, N. A., Bird, P. H. and Wallbridge, M. G. H. (1968): Inorg. Chem. 7, 1575.

Banister, A. J., Moore, L. F. and Padley, J. S. (1968): Inorganic Sulphur Chemistry (ed.: Nickless, G.). Elsevier, Amsterdam.

Bartell, L. S. (1960): J. Chem. Phys. 32, 832.

Bartell, L. S. (1972): Electron Diffraction by Gases. Physical Methods in Organic Chemistry (ed.: Weissberger, A. and Rossiter, B. W.). Interscience, New York.

Bartell, L. S. and Brockway, L. O. (1960): J. Chem. Phys. 32, 512.

Bartell, L. S. and Carroll, B. L. (1965a): J. Chem. Phys. 42, 1135.

Bartell, L. S. and Carroll, B. L. (1965b): J. Chem. Phys. 42, 3076.

Bartell, L. S. and Hirst, R. C. (1959): J. Chem. Phys. 31, 449.

Bartell, L. S., Kuchitsu, K. and Seip, H. M. (1974): Guide for the Publication of Experimental Gas-Phase Electron Diffraction Data and Derived Structural Results in the Primary Literature, to be published.

Bartell, L. S., Roth, E. A., Hollowell, C. D., Kuchitsu, K. and Young, J. E., Jr. (1965): J. Chem. Phys. 42, 2683.

Bastiansen, O. and Beagley, B. (1964): Acta Chem. Scand. 18, 2077.

Bauer, S. H. (1937): J. Am. Chem. Soc. 59, 1804.

Bauer, S. H. (1950): J. Am. Chem. Soc. 72, 622.

Bauer, S. H. (1970): Diffraction of Electrons by Gases. Physical Chemistry. Vol. 4 (ed.: Henderson, D.). Academic Press, New York.

Bauer, S. H. and Addison, C. C. (1960): Proc. Chem. Soc., 251.

Bauer, S. H., Beach, J. Y. and Simons, J. H. (1939): J. Am. Chem. Soc. 61, 19.

Bauer, S. H., Diner, R. M. and Porter, R. F. (1958): J. Chem. Phys. 29, 991.

Bauer, S. H., Ino, T. and Porter, R. F. (1960): J. Chem. Phys. 33, 685.

Bauer, S. H. and Porter, R. F. (1964): Metal Halide Vapors: Structures and Thermochemistry. Molten Salt Chemistry (ed.: Blander, M.). Interscience, New York.

Beach, J. Y. and Bauer, S. H. (1940): J. Am. Chem. Soc. 62, 3440.

Beagley, B., Cruickshank, D. W. J., Hewitt, T. G. and Haaland, A. (1967): Trans. Faraday Soc. **63**, 836.

Beagley, B., Cruickshank, D. W. J., Hewitt, T. G. and Jost, K. H. (1969): Trans. Faraday Soc. **65**, 1219.

Beagley, B., Cruickshank, D. W. J., Pinder, P. M., Robiette, A. G. and Sheldrick, G. M. (1969): Acta Cryst. **B25**, 737.

Beagley, B. and Hewitt, T. G. (1968): Trans. Faraday Soc. **64**, 2561.

Beagley, B., Schmidling, D. G. and Cruickshank, D. W. J. (1973): Acta Cryst. **B29**, 1499.

Beall, H. and Lipscomb, W. N. (1967): Inorg. Chem. **6**, 874.

Beattie, I. R. and Horder, J. R. (1969): J. Chem. Soc., 2655.

Beattie, I. R., Livingston, K. M. S., Ozin, G. A. and Reynolds, D. J. (1970): J. Chem. Soc. A., 449.

Beattie, I. R. and Ozin, G. A. (1970): J. Chem. Soc. A., 370.

Beaudet, R. A. and Poynter, R. L. (1965): J. Chem. Phys. **43**, 2166.

Beaudet, R. A. and Poynter, R. L. (1970): J. Chem. Phys. **53**, 1890.

Becker, E. D. and Pimentel, G. C. (1956): J. Chem. Phys. **25**, 224.

Begun, G. M. and Fletcher, W. H. (1960): J. Mol. Spectry. **4**, 388.

Benedict, W. S. and Plyler, E. K. (1957): Can. J. Phys. **35**, 1235.

Bennett, M. J., Cotton, F. A. and Foxman, B. M. (1968): Inorg. Chem. **7**, 1563.

Bentham, J. E. and Rankin, D. W. H. (1971): J. Organometal. Chem. **30**, C54.

Berkowitz, J. and Chupka, W. A. (1958): J. Chem. Phys. **29**, 653.

Berkowitz, J., Chupka, W. A. and Inghram, M. G. (1957): J. Chem. Phys. **27**, 85.

Berkowitz, J., Inghram, M. G. and Chupka, W. A. (1957): J. Chem. Phys. **26**, 842.

Berkowitz, J. and Marquart, J. R. (1962): J. Chem. Phys. **37**, 1853.

Berns, R. P., De Maria, G., Drowart, J. and Inghram, M. G. (1963): J. Chem. Phys. **38**, 1035.

Berry, R. S. and Klemperer, W. (1957): J. Chem. Phys. **26**, 724.

Bersuker, I. B. (1971): Stroenie i svoistva koordinatsionnykh soedinenii. Khimiya, Leningrad.

Bigelow, M. J. (1969): J. Chem. Educ. **46**, 495.

Bilton, M. S. an d Webster, M. (1972): J. Chem. Soc. A., 722.

Bird, P. H. and Churcill, M. R. (1967): Chem. Commun., 403.

Bjørseth, A., Drew, D. A., Marstokk, K. M. and Møllendal, H. (1972): J. Mol. Struct. **13**, 233.

Black, D. L. (1971): Dissertation, University of Michigan, Ann Arbor, Michigan.

Blank, J. S. (1964): Dissertation, Oregon State University, Corvallis, Oregon. Cited in McClelland, Gundersen and Hedberg (1972).

Boal, D. H., Briggs, G., Huber, H., Ozin, G. A., Robinson, E. A. and Vander Voet, A. (1971): Nature **231**, 174.

Boal, D. H. and Ozin, G. A. (1972): J. Chem. Soc. Dalton 1824.

Bohn, R. K. and Bauer, S. H. (1967): Inorg. Chem. **6**, 304.

Bohn, R. K. and Bohn, M. D. (1971): Inorg. Chem. **10**, 350.

Bohn, R. K. and Haaland, A. (1966): J. Organometal. Chem. **5**, 470.

Bonham, R. A. and Su, L. S. (1968): Second Austin Symposium on Gas Phase Molecular Structures. University of Texas, Austin, Texas.

Bor, Gy. (1966): The Chemistry of Metal Carbonyls and Their Derivatives. (In Hungarian.) Akadémiai Kiadó, Budapest.

Brandon, J. K. and Brown, I. D. (1968): Can. J. Chem. **46**, 933.

Brendhaugen, K., Haaland, A. and Novak, D. P. (1974): Acta Chem. Scand. **A28**, 45.

Brewer, L. and Lofgren, N. L. (1950): J. Am. Chem. Soc. **72**, 3038.

Brezgin, Yu. A. (1972): Dissertation, Moscow State University, Moscow.

Bridges, D. M., Holywell, G. C., Rankin, D. W. H. and Freeman, J. M. (1971): J. Organometal. Chem. **32**, 87.

Brittain, A. H., Cox, A. P. and Kuczkowski, R. L. (1969): Trans. Faraday Soc. **65**, 1963.

Brockway, L. O. and Anderson, J. S. (1937): Trans. Faraday Soc. **33**, 1233.

Brockway, L. O. and Cross, P. C. (1935): J. Chem. Phys. **3**, 828.

Brockway, L. O. and Davidson, N. R. (1941): J. Am. Chem. Soc. **63**, 3287.

Brockway, L. O., Ewens, R. V. C. and Lister, M. (1938): Trans. Faraday Soc. **34**, 1350.

Brom, J. M., Jr. and Franzen, H. F. (1971): J. Chem. Phys. **54**, 2874.

Brown, D. G., Drago, R. S. and Bolles, T. F. (1968): J. Am. Chem. Soc. **90**, 5706.

Brumbach, S. B. and Rosenblatt, G. M. (1972): J. Chem. Phys. **56**, 3110.

Brunvoll, J. (1965): J. Mol. Spectry. **15**, 386.

Brunvoll, J. (1967a): Private communication; cited in Seip (1967).

Brunvoll, J. (1967b): Acta Chem. Scand. **21**, 1390.

Brunvoll, J. and Cyvin, S. J. (1968): Acta Chem. Scand. **22**, 2709.

Brunvoll, J., Cyvin, S. J., Ewbank, J. D. and Schäfer, L. (1972): Acta Chem. Scand. **26**, 2161.

Brunvoll, J., Cyvin, S. J. and Schäfer, L. (1971): J. Organometal. Chem. **27**, 69.

Brunvoll, J., Cyvin, S. J. and Schäfer, L. (1972): J. Organometal. Chem. **36**, 143.

Bryan, P. S. and Kuczkowski, R. L. (1971): Inorg. Chem. **10**, 200.

Bryan, P. S. and Kuczkowski, R. L. (1972): Inorg. Chem. **11**, 553.

Burbank, R. D. (1965): Acta Cryst. **18**, 88.

Burg, A. B. and Schlesinger, H. I. (1940): J. Am. Chem. Soc. **62**, 3425.

Burns, J. H. and Gordon, E. K. (1966): Acta Cryst. **20**, 135.

Butaev, B. S. (1970): Thesis, Moscow State University, Moscow.

Büchler, A. and Berkowitz-Mattuck, J. B. (1967): Advances in High-Temperature Chemistry (ed.: Eyring, L.). Academic Press, New York.

Büchler, A., Blackburn, P. E. and Stauffer, J. L. (1966): J. Phys. Chem. **70**, 685.

Büchler, A. and Marram, E. P. (1963): J. Chem. Phys. **39**, 292.

Büchler, A., Stauffer, J. L. and Klemperer, W. (1964a): J. Am. Chem. Soc. **86**, 4544.

Büchler, A., Stauffer, J. L. and Klemperer, W. (1964b): J. Chem. Phys. **40**, 3471.

Büchler, A., Stauffer, J. L. and Klemperer, W. (1967): J. Chem. Phys. **46**, 605.

Byram, S. K., Fawcett, J. K., Nyberg, S. C. and O'Brien, R. J. (1970): Chem. Commun., 16.

Cardillo, M. J. and Bauer, S. H. (1969): Inorg. Chem. **8**, 2086.

Caron, A., Palenik, G. J., Goldish, E. and Donohue, J. (1964): Acta Cryst. **17**, 102.

16   241

Carroll, B. L. and Bartell, L. S. (1968): Inorg. Chem. **7**, 219.

Cartwright, B. S. and Robertson, J. H. (1966): Chem. Commun., 82.

Cheung, C. S. and Beaudet, R. A. (1971): Inorg. Chem. **10**, 1144.

Cheung, C. S., Beaudet, R. A. and Segal, G. A. (1970): J. Am. Chem. Soc. **92**, 4158.

Clark, A. H., Beagley, B. and Cruickshank, D. W. J. (1968): Chem. Commun., 14.

Claudine, F. B. (1965): Acta Cryst. **18**, 827.

Clementi, E. (1967a): J. Chem. Phys. **46**, 3851.

Clementi, E. (1967b): J. Chem. Phys. **47**, 2323.

Clippard, P. H., Hanson, J. C. and Taylor, R. C. (1971): J. Cryst. Mol. Struct. **1**, 363.

Cohen, E. A. and Beaudet, R. A. (1968): J. Chem. Phys. **48**, 1220.

Cohen, E. A. and Beaudet, R. A. (1973): Inorg. Chem. **12**, 1570.

Companion, A. L. (1972): J. Chem. Phys. **57**, 1807.

Cook, T. H. and Morgan, G. L. (1969): J. Am. Chem. Soc. **91**, 774.

Cook, T. H. and Morgan, G. L. (1970a): J. Am. Chem. Soc. **92**, 6487.

Cook, T. H. and Morgan, G. L. (1970b): J. Am. Chem. Soc. **92**, 6493.

Cornwell, C. D. (1950): J. Chem. Phys. **18**, 1118.

Costain, C. C. and Srivastava, G. P. (1961): J. Chem. Phys. **35**, 1903.

Costain, C. C. and Srivastava, G. P. (1964): J. Chem. Phys. **41**, 1620.

Cotton, F. A. (1970): Inorg. Chem. **9**, 2804.

Cotton, F. A., Dollase, W. A. and Wood, J. S. (1963): J. Am. Chem. Soc. **85**, 1543.

Cotton, F. A. and Mague, J. T. (1964): Inorg. Chem. **3**, 1402.

Cotton, F. A. and Reynolds, L. T. (1958): J. Am. Chem. Soc. **80**, 269.

Cotton, F. A. and Richardson, D. C. (1966): Inorg. Chem. **5**, 1851.

Cotton, F. A. and Wilkinson, G. (1972): Advanced Inorganic Chemistry. Third Edition. Interscience, New York.

Cotton, F. A. and Wing, R. M. (1965): Inorg. Chem. **4**, 867.

Cox, A. P. and Brittain, A. H. (1970): Trans. Faraday Soc. **66**, 557.

Cox, A. P. and Riveros, J. M. (1965): J. Chem. Phys. **42**, 3106.

Coyle, T., Lafferty, W. J. and Maki, A. G. (1967): cited in Kuchitsu (1968).

Craig, D. C. and Stephenson, N. C. (1968): Acta Cryst. **B24**, 1250.

Crasnier, F., Labarre, J.-F. and Leibovici, C. (1972): J. Mol. Struct. **14**, 405.

Cruickshank, D. W. J. (1961): J. Chem. Soc., 5486.

Cubicciotti, D. (1970): High Temp. Sci. **2**, 65.

Cyvin, B. N., Hargittai, M., Cyvin, S. J. and Hargittai, I. (1975): Acta Chim. (Budapest) **84**, 55.

Cyvin, S. J. (1968): Molecular Vibrations and Mean Square Amplitudes. Universitetsforlaget, Oslo and Elsevier, Amsterdam.

Cyvin, S. J. (1970): Z. anorg. allgem. Chem. **378**, 117.

Cyvin, S. J. (1973a): Private communication.

Cyvin, S. J. (1973b): Tidsskr. Kjemi, Bergv., Metallurgi **33**, nr. 3, 7.

Cyvin, S. J. and Andreassen, R. (1974): All India Symposia on Physics Education and Research. Kerala Physics Academy

Cyvin, S. J. and Brunvoll, J. (1969): J. Mol. Struct. **3**, 453.

Cyvin, S. J., Brunvoll, J. and Schäfer, L. (1971): J. Chem. Phys. **54**, 1517·

Cyvin, S. J. and Cyvin, B. N. (1971): Z. Naturforsch. **26a**, 901.
Cyvin, S. J., Cyvin, B. N. and Brunvoll, J. (1970): Acta Chem. Scand. **24**, 3420.
Cyvin, S. J., Cyvin, B. N. and Brunvoll, J. (1972): Adamantane — Part III: Normal Coordinate Analysis and Mean Amplitudes. Molecular Structures and Vibrations (ed.: Cyvin, S. J.). Elsevier, Amsterdam.
Cyvin, S. J., Cyvin, B. N., Brunvoll, J., Andersen, B. and Stølevik, R. (1967): Spectroscopic Studies in Connection with an Electron-Diffraction Investigation of Some Simple Molecules. Selected Topics in Structure Chemistry (ed.: Andersen, P., Bastiansen, O. and Furberg, S.). Universitetsforlaget, Oslo.
Cyvin, S. J., Cyvin, B. N. and Hargittai, I. (1974): J. Mol. Struct. **23**, 385
Cyvin, S. J., Cyvin, B. N., Rao, D. B. and Snelson, A. (1971): Z. anorg. allgem. Chem. **380**, 212.
Cyvin, S. J., Cyvin, B. N. and Snelson, A. (1971): J. Phys. Chem. **75**, 2609.
Cyvin, S. J. and Müller, A. (1971): Acta Chem. Scand. **25**, 1149.
Cyvin, S. J. and Vizi, B. (1969): Acta Chim. Acad. Sci. Hung. **59**, 85.
Dahl, L. F., Ishishi, E. and Rundle, R. E. (1957): J. Chem. Phys. **26**, 1750.
Dahl, L. F. and Rundle, R. E. (1963): Acta Cryst. **16**, 419.
Davies, N., Wallbridge, M. G. H., Smith, B. E. and James, B. D. (1973): J. C. S. Dalton Transactions, 162.
Davis, M. and Hanson, H. P. (1965): J. Phys. Chem. **69**, 3405.
Davis, M. and Hanson, H. P. (1967): J. Phys. Chem. **71**, 775.
Davis, M. I. and Speed, C. S. (1970): J. Organometal. Chem. **21**, 401.
Deever, W. R., Lory, E. R. and Ritter, D. M. (1969): Inorg. Chem. **8**, 1263.
Delaplane, R. G. and Ibers, J. A. (1969): Acta Cryst. **B25**, 2423.
Del Bene, J. E. and Pople, J. A. (1971): J. Chem. Phys. **55**, 2296.
Derissen, J. L. (1971a): J. Mol. Struct. **7**, 67.
Derissen, J. L. (1971b): J. Mol. Struct. **7**, 81.
Devarajan, V. and Cyvin, S. J. (1972): Acta Chem. Scand. **26**, 1.
Dewar, M. J. S. (1969): The Molecular Orbital Theory of Organic Chemistry. McGraw-Hill, New York.
Dinerman, C. E. and Ewing, G. E. (1970): J. Chem. Phys. **53**, 626.
Dodge, R. P. and Schomaker, V. (1965): Acta Cryst. **18**, 614.
Donohue, J. and Caron, A. (1964): Acta Cryst. **17**, 663.
Donohue, J. and Caron, A. (1966): J. Phys. Chem. **70**, 603.
Donohue, J. and Caron, A. (1967): J. Phys. Chem. **71**, 777.
Downs, A. J. (1968): Unusual Coordination Numbers. New Pathways in Inorganic Chemistry (ed.: Ebsworth, E. A. V., Maddock, A. G. and Sharpe, A. G.). University Press, Cambridge.
Drew, D. A., Gundersen, G. and Haaland, A. (1972): Acta Chem. Scand. **26**, 2147.
Drew, D. A. and Haaland, A. (1972a): Acta Cryst. **B28**, 3671.
Drew, D. A. and Haaland, A. (1972b): Chem. Commun., 1300.
Drew, D. A. and Haaland, A. (1972c): Acta Chem. Scand. **26**, 3079.
Drew, D. A. and Haaland, A. (1972d): Acta Chem. Scand. **26**, 3351.
Drew, D. A. and Haaland, A. (1973): Acta Chem. Scand. **27**, 3735.
Drew, D. A., Haaland, A. and Weidlein, J. (1973): Z. anorg. allgem. Chem. **398**, 241.

Drew, D. A. and Morgan, G. L. (1971): Unpublished results, cited in Drew and Haaland (1972c).

Dulmage, W. J. and Lipscomb, W. N. (1951): J. Am. Chem. Soc. **73**, 3539.

Dulmage, W. J. and Lipscomb, W. N. (1952): Acta Cryst. **5**, 260.

Dunks, G. B. and Hawthorne, M. F. (1968): J. Am. Chem. Soc. **90**, 7355.

Durig, J. R., Hudgens, B. A. and Odom, J. D. (1974): Inorg. Chem. **13**, 2306.

Durig, J. R., Li, Y. S., Carreira, L. A. and Odom, J. D. (1973): J. Am. Chem. Soc. **95**, 2491.

Durig, J. R., Li, Y. S. and Odom, J. D. (1973): J. Mol. Struct. **16**, 443.

Durig, J. R., Saunders, J. E. and Odom, J. D. (1971): J. Chem. Phys. **54**, 5285.

Durig, J. R., Thompson, J. W., Witt, J. D. and Odom, J. D. (1973): J. Chem. Phys. **58**, 5339.

Dyke, T. R., Howard, B. J. and Klemperer, W. (1972): J. Chem. Phys. **56**, 2442.

Ebsworth, E. A. V. (1963): Volatile Silicon Compounds. Pergamon Press. Oxford, London, etc.

Edgell, W. F., Wilson, W. E. and Summit, R. (1963): Spectrochim. Acta **19**, 863.

Edwards, A. J. (1964): J. Chem. Soc., 3714.

Egorova, N. M. and Rambidi, N. G. (1972): The Material Point Method in the Interpretation of Electron Diffraction Data. Molecular Structures and Vibrations (ed.: Cyvin, S. J.). Elsevier, Amsterdam.

Einstein, F. W. B., Gilbert, M. M. and Tuck, D. G. (1972): Inorg. Chem. **11**, 2832.

Eisenberg, R. and Ibers, J. A. (1965): Inorg. Chem. **4**, 773.

Eisenstadt, M., Rothberg, G. M. and Kusch, P. (1958): J. Chem. Phys. **29**, 797.

Eliezer, I. and Reger, A. (1972/73): Coord. Chem. Rev. **9**, 189.

Emerson, G. F., Ehrlich, K., Giering, W. P. and Lauterbur, P. C. (1966): J. Am. Chem. Soc. **88**, 3172.

Epstein, I. R., Koetzle, T. F., Stevens, R. M. and Lipscomb, W. N. (1970): J. Am. Chem. Soc. **92**, 7019.

Ewens, R. V. G. and Lister, M. (1939): Trans. Faraday Soc. **35**, 681.

Ezhov, Yu. S., Komarov, S. A. and Tolmachev, S. M. (1973): All-Union Conference on Khimiya paroobraznykh neorganicheskikh soedinenii i protsessov paroobrazovaniya. Minsk.

Ezhov, Yu. S., Tolmachev, S. M. and Rambidi, N. G. (1970): Zh. Strukt. Khim. **11**, 527.

Ezhov, Yu. S., Tolmachev, S. M. and Rambidi, N. G. (1972): Zh. Strukt. Khim. **13**, 972.

Ezhov, Yu. S., Tolmachev, S. M., Spiridonov, V. P. and Rambidi, N. G. (1968): Teplofiz. Vys. Temp. **6**, 68.

Fately, W. G., Bent, H. A. and Crawford, B., Jr. (1959): J. Chem. Phys. **31**, 204.

Fehler, T. P. and Koski, W. S. (1964): J. Am. Chem. Soc. **86**, 2733.

Ferguson, A. C. and Cornwell, C. D. (1970): J. Chem. Phys. **53**, 1851.

Fields, R., Germain, M. M., Haszeldine, R. N. and Wiggans, P. W. (1970): J. Chem. Soc. A., 1964; 1969.

Finch, A., Hyamas, I. and Steele, D. (1965): Spectrochim. Acta **21**, 1423.

Fischer, E. O. and Phab, W. (1952): Z. Naturforsch. **76,** 377.

Fischer, O. and Fritz, H. P. (1959): Advances in Inorganic Chemistry and Radio-chemistry. **1.** Academic Press, New York.

Flood E. (1974): J. Mol. Struct. **21,** 221.

Fritz, H. P. (1964): Advances in Organometallic Chemistry. **1.** Academic Press, New York.

Fritz, H. P. and Lüttke, W. (1959): Proc. Vth Intern. Conf. Coord. Chem. The Chemical Society, Spec. Publ. No. 13, London.

Fritz, H. P., Lüttke, W., Stammreich, H. and Forneris, R. (1959): Chem. Ber. **92,** 3246.

Frost, A. V., Akishin, P. A., Gurvich, L. V., Kurkchi, G. A. and Konstantinov, A. A. (1953): Vest. Moskov. Univ. Ser. Khim. No. 12, 85.

Ganis, P. and Allegra, G. (1963): cited in Ronova and Alekseev (1969).

Gantam, H. O. (1970): Indian J. Pure Appl. Phys. **8,** 713.

Gapotchenko, N. I., Alekseev, N. V., Anisimov, K. N., Kolobova, N. E. and Ronova, I. A. (1968): Zh. Strukt. Khim. **9,** 892.

Gapotchenko, N. I., Alekseev, N. V., Antonova, A. B., Anisimov, K. N., Kolobova, N. E., Ronova, I. A. and Struchkov, Yu. T. (1970): J. Organo-metal. Chem. **23,** 525.

Gapotchenko, N. I., Struchkov, Yu. T., Alekseev, N. V. and Ronova, I. A. (1973): Zh. Strukt. Khim. **14,** 419.

Gatehouse, B. M. and Leverett, P. (1969): J. Chem. Soc. A., 849.

Gayles, J. N. and Self, J. (1964): J. Chem. Phys. **40,** 3530.

Geller, S. (1960): J. Chem. Phys. **32,** 1569.

Geller, S. and Hoard, J. L. (1950): Acta Cryst. **3,** 121.

Geller, S. and Hoard, J. L. (1951): Acta Cryst. **4,** 399.

Gelus, M. and Kutzelnigg, W. (1973): Theor. Chim. Acta **28,** 103.

Gilbert, M. M., Gundersen, G. and Hedberg, K. (1972): J. Chem. Phys. **56,** 1691.

Gillespie, R. J. (1963): J. Chem. Educ. **40,** 295.

Gillespie, R. J. (1972): Molecular Geometry. Van Nostrand Reinhold Co., London.

Gillespie, R. J. and Nyholm, R. (1957): Quart. Rev. Chem. Soc. **11,** 339.

Goldfinger, P. and Verhaegen, G. (1969): J. Chem. Phys. **50,** 1467.

Gordon, S. and Campbell, C. (1955): Analyt. Chem. **27,** 1102.

Gordy, W. and Cook, R. L. (1970): Microwave Molecular Spectra. Inter-science, New York.

Gosling, K., McLaughlin, G. M., Sim, G. A. and Smith, J. D. (1970): Chem. Commun. 1617.

Grant, D. F., Killean, R. C. G. and Lawrence, J. L. (1969): Acta Cryst. **B25,** 377.

Grimley, R. T. (1967): Mass Spectrometry. The Characterization of High-Temperature Vapors (ed.: Margrave, J. L.). Wiley, New York.

Gropen, O. and Haaland, A. (1973): Acta Chem. Scand. **27,** 521.

Groth, P. (1963): Acta Chem. Scand. **17,** 2419.

Gundersen, G., Haugen, T. and Haaland, A. (1973): J. Organometal. Chem. **54,** 77.

Gundersen, G., Hedberg, L. and Hedberg, K. (1973): J. Chem. Phys. **59,** 3777.

Haaland, A. (1965): Acta Chem. Scand. **19,** 41.

Haaland, A. (1968): Acta Chem. Scand. **22,** 3030.

Haaland, A. (1969): Studies on the Structures of Organometallic Compounds. Universitetsforlaget, Oslo.

Haaland, A. (1973): Private communication.

Haaland, A., Lusztyk, J., Novak, D. P., Brunvoll, J. and Starowieyski, K. B. (1974): J. C. S. Chem. Commun., 54.

Haaland, A. and Nilsson, J. E. (1968a): Chem. Commun., 88.

Haaland, A. and Nilsson, J. E. (1968b): Acta Chem. Scand. **22,** 2653.

Haaland, A. and Novak, D. P. (1974): Acta Chem. Scand. **A28,** 153.

Haaland, A. and Weidlein, J. (1972): J. Organometal. Chem. **40,** 29.

Hallam, H. E. (ed.) (1973): Vibrational Spectroscopy of Trapped Species. Infrared and Raman Studies of Matrix-Isolated Molecules, Radicals and Ions. Wiley — Interscience, London.

Hampson, G. C. and Stosick, A. J. (1938): J. Am. Chem. Soc. **60,** 1814.

Handy, L. B., Treichel, P. M., Dahl, L. F. and Hayter, R. G. (1966): J. Am. Chem. Soc. **88,** 336.

Hanic, F. and Subrtová, V. (1969): Acta Cryst. **B25,** 405.

Hansen, K. W. and Bartell, L. S. (1965): Inorg. Chem. **4,** 1775.

Hanson, A. W. (1962): Acta Cryst. **15,** 930.

Hardgrave, G. L. and Templeton, D. H. (1959): Acta Cryst. **12,** 28.

Hardy, C. J. and Field, B. O. (1963): J. Chem. Soc., 5130.

Hargittai, I. (1974): The Electron Diffraction Internuclear Distance. (In Hungarian.) A kémia újabb eredményei (Advances in Chemistry), Vol. 21 (ed.: Csákvári, B.). Akadémiai Kiadó, Budapest.

Hargittai, I. and Hargittai, M. (1974): J. Chem. Phys. **60,** 2563.

Hargittai, I., Hargittai, M., Spiridonov, V. P. and Erokhin, E. V. (1971): J. Mol. Struct. **8,** 31.

Hargittai, I. and Hedberg, K. (1972): Adamantane — Part I: Reinvestigation of the Molecular Structure by Electron Diffraction. Molecular Structures and Vibrations (ed.: Cyvin, S. J.). Elsevier, Amsterdam.

Hargittai, I. and Mijlhoff, F. C. (1973): J. Mol. Struct. **16,** 69.

Hargittai, I., Tremmel, J. and Schultz, Gy. (1975): J. Mol. Struct. **26,** 116.

Hargittai, M. (1970): Dissertation, Budapest.

Hargittai, M., Hargittai, I., Cyvin, B. N. and Cyvin, S. J. (1974): All India Symposia on Physics Education and Research. Kerala Physics Academy.

Hargittai, M., Hargittai, I. and Spiridonov, V. P. (1973): J. C. S. Chem. Commun., 750.

Hargittai, M., Hargittai, I. and Spiridonov, V. P. (1976): J. Mol. Struct. **30,** 31.

Hargittai, M., Hargittai, I., Spiridonov, V. P., Pelissier, M. and Labarre, J.-F. (1975): J. Mol. Struct. **24,** 27.

Hargittai, M., Hargittai, I., Tamás, J., Bihari, M., Szpiridonov, V. P. and Ivanov, A. A. (1974): Magy. Kém. Foly. **80,** 442.

Hastie, J. W., Hauge, R. and Margrave, J. L. (1968): J. Phys. Chem. **72,** 4492.

Hawkins, N. J., Mattraw, H. C., Sabol, W. W. and Carpenter, D. R. (1955): J. Chem. Phys. **23,** 2422.

Hedberg, K. (1973): Private communication.

Hedberg, K. (1974): Private communication.

246

Hedberg, K. and Iwasaki, M. (1960): J. Chem. Phys. 36, 589.
Hedberg, K., Jones, M. E. and Schomaker, V. (1951): J. Am. Chem. Soc. 73, 3538.
Hedberg, K., Jones, M. E. and Schomaker, V. (1952): Proc. Natl. Acad. Sci. U. S. 38, 679.
Hedberg, K. and Schomaker, V. (1951): J. Am. Chem. Soc. 73, 1482.
Hedberg, K. and Stosick, A. J. (1952): J. Am. Chem. Soc. 74, 954.
Hedberg, L. and Hedberg, K. (1970): J. Chem. Phys. 53, 1228.
Hersch, O. L. (1963): Dissertation. University of Michigan, Ann Arbor, Michigan. Cited in Kuczkowski (1968).
Hess, H. (1969): Acta Cryst. B25, 2338.
Higginbotham, H. K. and Bartell, L. S. (1965): J. Chem. Phys. 42, 1131.
Hildenbrand, D. L., Hall, W. F. and Potter, N. D. (1963): J. Chem. Phys. 39, 296.
Hisatsune, I. C., Devlin, J. P. and Wada, Y. (1960): J. Chem. Phys. 33, 714.
Hoard, J. L., Geller, S. and Cashin, W. M. (1951): Acta Cryst. 4, 396.
Hoard, J. L., Geller, S. and Owen, T. B. (1951): Acta Cryst. 4, 405.
Hoard, J. L., Owen, T. B., Buzzell, A. and Salmon, O. N. (1950): Acta Cryst. 3, 130.
Hoffmann, R. and Lipscomb, W. N. (1962): J. Chem. Phys. 36, 3489.
Holtzberg, F., Post, B. and Fankuchen, I. (1953): Acta Cryst. 6, 127.
Hrostowski, H. J. and Myers, R. J. (1954): J. Chem. Phys. 22, 262.
Hrostowski, H. J., Myers, R. J. and Pimentel, G. C. (1952): J. Chem. Phys. 20, 518.
Huffman, J. C. and Streib, W. E. (1971): Chem. Commun., 911.
Ionov, S. P. and Ionova, G. V. (1970): Izvest. Akad. Nauk S. S. S. R. Ser. Khim., 2836.
Ishchenko, A. A., Spiridonov, V. P. and Zasorin, E. Z. (1974): Izv. vysshikh uchebnykh zav. Khim. i Khim. Tekhnol. 17, 138.
Ishchenko, A. A., Zasorin, E. Z., Spiridonov, V. P., Bersuker, I. B. and Budnikov, S. S. (1974): Fifth All-Union Conference on Fizicheskie i matematicheskie metody v koordinatsionnoi khimii. Kishinev.
Ivanov, A. A., Spiridonov, V. P., Erokhin, E. V. and Levitskii, V. A. (1973): Zh. Fiz. Khim. 47, 3030.
Janzen, J. and Bartell, L. S. (1969): J. Chem. Phys. 50, 3611.
Jarvis, J. A. J., Mais, R. H. B. and Owston, P. G. (1968): J. Chem. Soc. A., 1473.
Jeffrey, G. A., Parry, G. S. and Mozzi, R. L. (1956): J. Chem. Phys. 25, 1024.
Jellinek, F. (1960): Nature 187, 871.
Jellinek, F. (1963): J. Organometal. Chem. 1, 43.
Jones, D. S. and Lipscomb, W. N. (1969): J. Chem. Phys. 51, 3133.
Jones, L. H. (1963): Spectrochim. Acta 19, 329.
Jones, M. E., Hedberg, K. and Schomaker, V. (1953): J. Am. Chem. Soc. 75, 4116.
Kaiser, E. W., Falconer, W. E. and Klemperer, W. (1972): J. Chem. Phys. 56, 5392.
Kaluski, Z. L. and Struchkov, Yu. T. (1965): Zh. Strukt. Khim. 6, 316.
Kaluski, Z. L., Struchkov, Yu. T. and Avoyan, R. L. (1964): Zh. Strukt. Khim. 5, 743.

Karle, J. (1973): Electron Diffraction. Determination of Organic Structures by Physical Methods, Vol. 5. Academic Press, New York.

Kasper, J. S., Lucht, C. M. and Harker, D. (1950): Acta Cryst. **3**, 436.

Kawai, K. and Murata, H. (1960): Bull. Chem. Soc. Japan **33**, 1008.

Kay, M. I., Frazer, B. C. and Almodovar, I. (1964): J. Chem. Phys. **40**, 504.

Kazenas, E. K. and Tsvetkov, Yu. V. (1967): Zh. Fiz. Khim. **41**, 3112.

Kealy, T. J. and Pauson, P. L. (1951): Nature **168**, 1039.

Keeling, R. O. (1957): Acta Cryst. **10**, 209.

Keneshea, F. J. and Cubicciotti, D. (1965): J. Phys. Chem. **69**, 3910.

Keneshea, F. J. and Cubicciotti, D. (1967): J. Phys. Chem. **71**, 1958.

Keulen, E. and Jellinek, F. (1966): J. Organometal. Chem. **5**, 490.

Khodchenkov, A. N. (1965): Dissertation. Moscow State University, Moscow.

Khodchenkov, A. N., Spiridonov, V. P. and Akishin, P. A. (1965): Zh. Strukt. Khim. **6**, 765.

Kihlborg, L. and Gebert, E. (1970): Acta Cryst. **B26**, 1020.

Kiindig, P., Moskovits, M. and Ozin, G. A. (1972): J. Mol. Struct. **14**, 137.

Kishida, S. and Nakamoto, K. (1964): J. Chem. Phys. **41**, 1558.

Kittelberger, J. S. and Hornig, D. F. (1967): J. Chem. Phys. **46**, 3099.

Klemperer, W. (1956): J. Chem. Phys. **24**, 353.

Klemperer, W. and Rice, S. A. (1957): J. Chem. Phys. **26**, 618.

Koetzle, T. F. and Lipscomb, W. N. (1970): Inorg. Chem. **9**, 2743.

Kollman, P. A. and Allen, L. C. (1970): J. Chem. Phys. **52**, 5085.

Konaka, S., Ito, T. and Morino, Y. (1966): Bull. Chem. Soc. Japan **39**, 1146.

Konaka, S. and Kimura, M. (1973): Bull. Chem. Soc. Japan **46**, 413.

Konaka, S., Murata, Y., Kuchitsu, K. and Morino, Y. (1966): Bull. Chem. Soc. Japan **39**, 1134.

Kools, F. X. N. M., Koster, A. S. and Rieck, G. D. (1970): Acta Cryst. **B26**, 1974.

Koster, A. S., Kools, F. X. N. M. and Rieck, G. D. (1969): Acta Cryst. **B25**, 1704.

Kovar, R. A. and Morgan, G. L. (1969): Inorg. Chem. **8**, 1099.

Kuchitsu, K. (1966): J. Chem. Phys. **44**, 906.

Kuchitsu, K. (1968): J. Chem. Phys. **49**, 4456

Kuchitsu, K (1971): Bull. Chem. Soc. Japan **44**, 96.

Kuchitsu, K. (1972): Gas Electron Diffraction. MTP International Review of Science. Physical Chemistry Series One, Vol. 2. Molecular Structure and Properties (ed.: Allen, G.). Medical and Technical Publ. Co., Oxford.

Kuchitsu, K. and Cyvin, S. J. (1972): Representation and Experimental Determination of the Geometry of Free Molecules. Molecular Structures and Vibrations (ed.: Cyvin, S. J.). Elsevier, Amsterdam.

Kuchitsu, K., Guillory, J. P. and Bartell, L. S. (1968): J. Chem. Phys. **49**, 2488.

Kuchitsu, K. and Konaka, S. (1966): J. Chem. Phys. **45**, 4342.

Kuchitsu, K., Shibata, S., Yokozeki, A. and Matsumura, C. (1971): Inorg. Chem. **10**, 2584.

Kuczkowski, R. L. (1965): J. Am. Chem. Soc. **87**, 5259.

Kuczkowski, R. L. (1968): J. Am. Chem. Soc. **90**, 1705.

Kuczkowski, R. L. and Lide, D. R. (1967): J. Chem. Phys. **46**, 357.

Kuczkowski, R. L., Lide, D. R. and Krisher, L. C. (1966): J. Chem. Phys. **44**, 3131.

Labarre, J.-F. and Leibovici, C. (1972): Int. J. Quantum Chem. **6**, 625.

La Placa, S. J., Hamilton, W. C., Ibers, J. A. and Davison, A. (1969): Inorg. Chem. **8**, 1928.

Laubengayer, A. W. and Lengnick, G. F. (1966): Inorg. Chem. **5**, 503.

LaVilla, R. E. and Bauer, S. H. (1963): J. Am. Chem. Soc. **85**, 3597.

Le Carpentier, J.-M. and Weiss, R. (1972): Acta Cryst. **B28**, 1437.

Lesiecki, M. L. and Shirk, J. S. (1972): J. Chem. Phys. **56**, 4171.

Levison, K. A. and Perkins, P. G. (1970): Theor. Chim. Acta **17**, 1.

Li, Y. S., Chen, M. M. and Durig, J. R. (1972): J. Mol. Struct. **14**, 261.

Lide, D. R. (1957): J. Chem. Phys. **27**, 343.

Lide, D. R. (1959): Spetrochim. Acta **15**, 473.

Lide, D. R. (1960): J. Chem. Phys. **32**, 1570.

Lide, D. R. (1969): Chemical Information from Microwave Spectroscopy. Survey of Progress in Chemistry, **5**, Academic Press, New York.

Lide, D. R., Cahill, P. and Gold, L. P. (1964): J. Chem. Phys. **40**, 156.

Lide, D. R. and Kuczkowski, R. L. (1967): J. Chem. Phys. **46**, 4768.

Lide, D. R. and Mann, D. E. (1958): J. Chem. Phys. **29**, 914.

Lide, D. R. and Matsumura, C. (1969): J. Chem. Phys. **50**, 3080.

Lide, D. R., Taft, R. W. and Love, P. (1959): J. Chem. Phys. **31**, 561.

Liebman, J. F. (1974): J. Chem. Phys. **60**, 2944.

Liehr, A. D. (1963): J. Phys. Chem. **67**, 471.

Linevsky, M. J. (1961): J. Chem. Phys. **34**, 587.

Lippincott, E. R. and Nelson, R. D. (1958): Spectrochim. Acta **10**, 307.

Lipscomb, W. N. (1963): Boron Hydrides. Benjamin, New York.

Lipscomb, W. N., Wang, F. E., May, W. R. and Lippert, E., Jr. (1961): Acta Cryst. **14**, 1100.

Lister, M. and Sutton, L. E. (1941): Trans. Faraday Soc. **37**, 393.

Litzow, M. R. (1973): in Litzow, M. R. and Spalding, T. R.: Mass Spectrometry of Inorganic and Organometallic Compounds. Physical Inorganic Chemistry. Monograph 2 (ed.: Lappert, M. F.). Elsevier, Amsterdam.

Long, L. H. (1972): Recent Studies of Diborane. Progress in Inorganic Chemistry **15**, 1.

Lopatko, O. Ya., Klimenko, N. M. and Dyatkina, M. E. (1972): Zh. Strukt. Khim. **13**, 1128.

Lory, E. R., Porter, R. F. and Bauer, S. H. (1971): Inorg. Chem. **10**, 1072.

Lovas, F. J., Tiemann, E. and Johnson, D. R. (1974): J. Chem. Phys. **60**, 5005.

Lu, C. S. and Donohue, J. (1944): J. Am. Chem. Soc. **66**, 818.

Macdonald, A. C. and Trotter, J. (1964): Acta Cryst. **17**, 872.

Mach, K. (1965): Collect. Czech. Commun. **30**, 277.

Mais, R. H. B., Owston, P. G. and Thompson, D. T. (1967): J Chem. Soc. A., 1735.

Mais, R. H. B., Owston, P. G., Thompson, D. T. and Wood, A. M. (1967): J. Chem. Soc. A., 1744.

Mandel, M. and Barrett, R. H. (1955): Phys. Rev. **99**, 666.

Mann, G., Haaland, A. and Weidlein, J. (1973): Z. anorg. allgem. Chem. **398**, 231.

Margrave, J. L. (Symposium Chairman) (1968): Mass Spectrometry in Inorganic Chemistry. Advances in Chemistry Series, Vol. 72 (ed.: Gould, R. F.). American Chemical Society Publications, Washington, D. C.

Marriott, J. C., Salthouse, J. A. and Ware, M. J. (1970): Chem. Commun., 595.

Marynick, D. S. and Lipscomb, W. N. (1971): J. Am. Chem. Soc. **93**, 2322.

Marynick, D. S. and Lipscomb, W. N. (1973): J. Am. Chem. Soc. **95**, 7244.

Mastryukov, V. S., Dorofeeva, O. V., Vilkov, L. V., Zhigach, A. F., Laptev, V. T. and Petrunin, A. B. (1973): J. C. S. Chem. Commun., 276.

Mastryukov, V. S., Vilkov, L. V., Zhigach, A. F. and Siryatskaya, V. N. (1969): Zh. Strukt. Khim. **10**, 136.

Mastryukov, V. S., Vilkov, L. V., Zhigach, A. F. and Siryatskaya, V. N. (1971): Zh. Strukt. Khim. **12**, 1081.

Maxwell, L. R., Hendricks, S. B. and Deming, L. S. (1937): J. Chem. Phys. **5**, 626.

McAdam, A., Francis, J. N. and Ibers, J. A. (1971a): J. Organometal. Chem. **29**, 131.

McAdam, A., Francis, J. N. and Ibers, J. A. (1971b): J. Organometal. Chem. **29**, 149.

McClelland, B. W., Gundersen, G. and Hedberg, K. (1972): J. Chem. Phys. **56**, 4541.

McCory, L. D., Paule, R. C. and Margrave, J. L. (1963): J. Phys. Chem. **67**, 1086.

McDonald, J. D. and Margrave, J. L. (1968): J. Inorg. Nucl. Chem. **30**, 665.

McDonald, T. R. R. and McDonald, W. S. (1972): Acta Cryst. **B28**, 1619.

McDonald, W. S. and Cruickshank, D. W. J. (1967): Acta Cryst. **22**, 48.

McGandy, E. L. (1961): Dissertation, Boston. Cited in Bryan and Kuczkowski (1972).

McKown, G. L. and Beaudet, R. A. (1971): Inorg. Chem. **10**, 1350.

McNeill, E. A., Gallaher, K. L., Scholer, F. R. and Bauer, S. H. (1973): Inorg. Chem. **12**, 2108.

Messmer, G. G. and Amma, E. L. (1966): Inorg. Chem. **5**, 1775.

Mijlhoff, F. C. (1964): Some Physical Properties and Structure of Selenium Trioxide. Dissertation, Amsterdam.

Mijlhoff, F. C. (1965a): Acta Cryst. **18**, 795.

Mijlhoff, F. C. (1965b): Rec. Trav. Chim. **84**, 74.

Millen, D. J. and Morton, J. R. (1960): J. Chem. Soc., 1523.

Miller, R. and Kusch, P. (1956): J. Chem. Phys. **25**, 860.

Miller, S. A., Tebboth, J. A. and Tremaine, J. F. (1952): J. Chem. Soc., 632.

Millner, T. (1955): Magy. Tud. Akad. Műszaki Tud. Oszt. Közl. **16**, 99.

Millner, T. and Neugebauer, J. (1949): Nature **163**, 601.

Milne, T. A., Klein, H. M. and Cubicciotti, D. (1958): J. Chem. Phys. **28**, 718.

Mole, T. and Jeffry, E. A. (1972): Organoaluminium Compounds. Elsevier, Amsterdam.

Morino, Y., Kuchitsu, K. and Moritani, T. (1969): Inorg. Chem. **8**, 867.

Morino, Y., Kuchitsu, K. and Yamamoto, S. (1968): Spectrochim. Acta **24A**, 335.

Morino, Y. and Uehara, U. (1966): J. Chem. Phys. **45**, 4543.

Morino, Y., Ukaji, T. and Ito, T. (1966a): Bull. Chem. Soc. Japan **39**, 64.

Morino, Y., Ukaji, T. and Ito, T. (1966b): Bull. Chem. Soc. Japan **39**, 71.
Moritani, T., Kuchitsu, K. and Morino, Y. (1971): Inorg. Chem. **10**, 344.
Møllendal, H. (1973): Private communication.
Muetterties, E. L. (1965): Inorg. Chem. **4**, 769.
Muetterties, E. L. (1970): Accounts Chem. Res. **3**, 266.
Muetterties, E. L. (1972): Stereochemical Non-rigidity. In MTP International
    Review of Science. Reaction Mechanisms in Inorganic Chemistry. Inorganic
    Chemistry Series One, Vol. 9 (Volume ed.: Tobe, M. L.). Medical and Tech-
    nical Publ. Co., Oxford.
Mulliken, R. S. and Person, W. B. (1969): Molecular Complexes. A Lecture
    and Reprint Volume. Wiley-Interscience, New York.
Nahringbauer, I. (1970): Acta Chem. Scand. **24**, 453.
Nakamoto, K. and Kishida, S. (1964): J. Chem. Phys. **41**, 1554.
Narten, A. H. (1972): J. Chem. Phys. **56**, 1905.
Ngai, L. H., Stafford, F. E. and Schäfer, L. (1969): J. Am. Chem. Soc. **91**, 48.
Nibler, J. W. (1972): J. Am. Chem. Soc. **94**, 3349.
Nibler, J. W. and Dyke, T. (1970): J. Am. Chem. Soc. **92**, 2920.
Nibler, J. W. and McNabb, J. (1969): Chem. Commun., 134.
Nibler, J. W., Shriver, D. F. and Cook, T. H. (1971): J. Chem. Phys. **54**, 5257.
Nielsen, A. H. (1954): J. Chem. Phys. **22**, 659.
Nimon, L. A., Seshadri, K. S., Taylor, R. C. and White, D. (1970): J. Chem.
    Phys. **53**, 2416.
Nixon, J. F. (1973): Endeavour **32**, No. 115, 19.
Nordman, C. E. (1960): Acta Cryst. **13**, 535.
Nordman, C. E. and Lipscomb, W. N. (1953a): J. Am. Chem. Soc. **75**, 4116.
Nordman, C. E. and Lipscomb, W. N. (1953b): J. Chem. Phys. **21**, 1856.
Novick, S. E., Howard, B. J. and Klemperer, W. (1972): J. Chem. Phys. **57**, 5619.
Novick, S. E., Howard, B. J. and Klemperer, W. (1974): J. Chem. Phys. **60**,
    2945.
Novikov, G. I. and Gavryuchenkov, F. G. (1967): Uspekhi Khim. **36**, 399.
Oberhammer, H. and Brune, H. A. (1969): Z. Naturforsch. **24a**, 607.
Odom, J. D., Saunders, J. E. and Durig, J. R. (1972): J. Chem. Phys. **56**,
    1643.
Olie, K. and Mijlhoff, F. C. (1969): Acta Cryst. **B25**, 974.
Onak, T., Dunks, G. B., Beaudet, R. A. and Poynter, R. L. (1966): J. Am.
    Chem. Soc. **88**, 4622.
Onak, T., Gerhart, F. J. and Williams, R. E. (1963): J. Am. Chem. Soc. **85**,
    3378.
Onak, T. P. and Wong, G. T. F. (1970): J. Am. Chem. Soc. **92**, 5226.
Otake, M., Matsumura, C. and Morino, Y. (1968): J. Mol. Spectry. **28**, 316.
Ozin, G. A. (1971): The SPEX Speaker **16**, 1.
Ozin, G. A. (1973): Matrix Isolation Laser Raman Spectroscopy, see Hallam
    (1973).
Ozin, G. A. and Vander Voet, A. (1971): J. Mol. Struct. **10**, 173.
Palmer, K. J. and Elliott, N. (1938): J. Am. Chem. Soc. **60**, 1852.
Parry, R. W. (1969): Coordination Compounds Containing Fluorophosphine
    Ligands. Coordination Chemistry (ed.: Kirchener, S.). Plenum Press, New
    York.

251

Pascard, R. and Pascard-Billy, C. (1965): Acta Cryst. **18**, 830.

Pasinski, J. P. and Beaudet, R. A. (1973): J. C. S. Chem. Commun., 928.

Pasinski, J. P. and Kuczkowski, R. L. (1971): J. Chem. Phys. **54**, 1903.

Pauling, L. (1960): The Nature of the Chemical Bond. Cornell University Press, Ithaca.

Pauling, L. and Brockway, L. O. (1934): Proc. Nat. Acad. Sci. **20**, 336.

Payne, D. S., Mokuolu, J. A. A. and Speakman, J. C. (1965): Chem. Commun., 599.

Pelissier, M., Labarre, J.-F., Vilkov, L. V., Golubinsky, A. V. and Mastryukov, V. S. (1974): J. Chim. Phys. **71**, 702.

Perov, P. A. (1973): Dissertation, Moscow State University, Moscow.

Pilar, F. L. (1968): Elementary Quantum Chemistry. McGraw-Hill, New York.

Plato, V., Hartford, W. D. and Hedberg, K. (1970): J. Chem. Phys. **53**, 3488.

Plato, V. and Hedberg, K. (1970): Inorg. Chem. **10**, 590.

Pople, J. A. and Beveridge, D. L. (1970): Approximate Molecular Orbital Theory. McGraw-Hill, New York.

Porter, R. F., Schoomaker, R. C. and Addison, C. C. (1959): Proc. Chem. Soc., 11.

Porter, R. F. and Zeller, E. E. (1960): J. Chem. Phys. **33**, 858.

Price, W. C. (1947): J. Chem. Phys. **15**, 614.

Price, W. C. (1948): J. Chem. Phys. **16**, 894.

Price, W. C. (1949): J. Chem. Phys. **17**, 1044.

Pugh, A. C. P. and Barrow, R. F. (1958): Trans. Faraday Soc. **54**, 671.

Pulay, P. (1969): Mol. Phys. **17**, 197.

Rahman, A., Fowler, R. H. and Narten, A. H. (1972): J. Chem. Phys. **57**, 3010.

Rambidi, N. G. (1973): All-Union Conference on Khimiya paroobraznykh neorganicheskikh soedinenii i protsessov paroobrazovaniya. Minsk.

Rambidi, N. G. and Spiridonov, V. P. (1964): Teplofiz. Vys. Temp. **2**, 280.

Rambidi, N. G. and Zasorin, E. Z. (1964): Teplofiz. Vys. Temp. **2**, 705.

Reichert, U. V. and Hartmann, H. (1972): Z. Naturforsch. **27a**, 983.

Renes, P. A. and MacGillavry, C. H. (1945): Rec. Trav. Chim. **64**, 275.

Rice, S. A. and Klemperer, W. (1957a): J. Chem. Phys. **27**, 573.

Rice, S. A. and Klemperer, W. (1957b): J. Chem. Phys. **27**, 643.

Rinke, K. and Schäfer, H. (1965): Angew. Chem. **77**, 131.

Robiette, A. G., Sheldrick, G. M. and Simpson, R. N. F. (1968): Chem. Commun., 506.

Robiette, A. G., Sheldrick, G. M. and Simpson, R. N. F. (1969): J. Mol. Struct. **4**, 221.

Robiette, A. G., Sheldrick, G. M., Simpson, R. N. F., Aylett, B. J. and Campbell, J. A. (1968): J. Organometal. Chem. **14**, 279.

Robinson, W. T. and Ibers, J. A. (1967): Inorg. Chem. **6**, 1208.

Roddatis, N. M., Tolmachev, S. M., Ugarov, V. V., Ezhov, Yu. S. and Rambidi, N. G. (1974): Fifth Austin Symposium on Gas Phase Molecular Structure. Austin, Texas.

Romanov, G. V. and Spiridonov, V. P. (1968): Vest. Moskov. Univ. Ser. Khim. No. 5, 7.

Ronova, I. A. and Alekseev, N. V. (1967): Dokl. Akad. Nauk S. S. S. R. **174**, 614.

252

Ronova, I. A. and Alekseev, N. V. (1969): Dokl. Akad. Nauk S. S. S. R. **185**, 1303.

Ronova, I. A., Alekseev, N. V., Gapotchenko, N. I. and Struchkov, Yu. T. (1970a): J. Organometal. Chem. **25**, 149.

Ronova, I. A., Alekseev, N. V., Gapotchenko, N. I. and Struchkov, Yu. T. (1970b): Zh. Strukt. Khim. **11**, 584.

Ronova, I. A., Bochvar, D. A., Chistjakov, A. L., Struchkov, Yu. T. and Alekseev, N. V. (1969): J. Organometal. Chem. **18**, 337.

Rudolph, R. W. and Parry, R. W. (1967): J. Am. Chem. Soc. **89**, 1621.

Rudolph, R. W., Parry, R. W. and Farran, C. F. (1966): Inorg. Chem. **5**, 723.

Rupp, J. J. and Shriver, D. F. (1967): Inorg. Chem. **6**, 755.

Ryan, R. R. and Hedberg, K. (1969): J. Chem. Phys. **50**, 4986.

Saito, S. (1963): J. Mol. Spectry. **30**, 1.

Sass, R. L., Vidale, R. and Donohue, J. (1957): Acta Cryst. **10**, 567.

Schäfer, L., Brunvoll, J. and Cyvin, S. J. (1972): J. Mol. Struct. **11**, 459.

Schäfer, L., Ewbank, J. D., Cyvin, S. J. and Brunvoll, J. (1972): J. Mol. Struct. **14**, 185.

Schäfer, L., Southern, J. F. and Cyvin, S. J. (1971): Spectrochim. Acta **27A**, 1083.

Schäfer, L., Southern, J. F., Cyvin, S. J. and Brunvoll, J. (1970): J. Organometal. Chem. **24**, C13.

Schirdewahn, H. G. (1965): Dissertation, Freiburg.

Schomaker, V. and Stevenson, D. P. (1941): J. Am. Chem. Soc. **63**, 37.

Schroeder, H., Heying, T. L. and Reiner, J. R. (1963): Inorg. Chem. **2**, 1092.

Seibold, E. A. and Sutton, L. E. (1955): J. Chem. Phys. **23**, 1967.

Seip, H. M. (1967): Studies on the Failure of the First Born Approximation in Electron Diffraction. Selected Topics in Structure Chemistry (ed.: Andersen, P., Bastiansen, O. and Furberg, S.). Universitetsforlaget, Oslo.

Seip, H. M. (1973): Theory; Accuracy. Molecular Structure by Diffraction Methods (senior reporters: Sim, G. A. and Sutton, L. E.). Specialist Periodical Reports, The Chemical Society, London.

Seip, H. M. and Seip, R. (1970): Acta Chem. Scand. **24**, 3431.

Seip, R. (1972): Acta Chem. Scand. **26**, 1966.

Selivanov, G. K. and Mal'tsev, A. A. (1973): Zh. Strukt. Khim. **14**, 943.

Semenenko, K. N., Lobkovskii, E. B., Bulichev, B. M., Golubinskii, A. V., Vilkov, L. V. and Mastryukov, V. S. (1973): XI All-Union Conference on Khimii kompleksnykh soedinenii, Alma Ata.

Semenenko, K. N., Lobkovskii, E. B. and Dorosinskii, A. L. (1972): Zh. Strukt. Khim. **13**, 743.

Semenov, G. A. (1970): cited in Roddatis, Tolmachev, Ugarov, Ezhov and Rambidi (1974).

Seshadri, K. S., Nimon, L. A. and White, D. (1969): J. Mol. Spectry. **30**, 128.

Severinsson, R. (1955): Dissertation, University of Oslo, Oslo. Cited in McClelland, Gundersen and Hedberg (1972).

Shapiro, I., Keilin, B., Williams, R. W. and Good, C. D. (1963): J. Am. Chem Soc. **85**, 3167.

Shen, Q. (1973): Dissertation, Oregon State University, Corvallis, Oregon.

Shibata, S. (1970): Acta Chem. Scand. **24**, 705.

253

Shibata, S., Bartell, L. S. and Gavin, R. M., Jr. (1964): J. Chem. Phys. **41**, 717.

Shol'ts, V. B. and Sidorov, L. N. (1972): Vest. Moskov. Univ. Ser. Khim. No. 4. 371.

Shriver, D. F. and Swanson, B. (1971): Inorg. Chem. **10**, 1354.

Sidgwick, N. V. and Powell, H. E. (1940): Proc. Roy. Soc. **A176**, 153.

Sidorov, L. N., Erokhin, E. V., Akishin, P. A. and Kolosov, E. N. (1967): Dokl. Akad. Nauk S. S. S. R. **173**, 370.

Sidorov, L. N. and Kolosov, E. N. (1968): Zh. Fiz. Khim. **42**, 2617.

Silbiger, G. and Bauer, S. H. (1946): J. Am. Chem. Soc. **68**, 312.

Sinke, E. J., Pressley, G. A., Jr., Baylis, A. B. and Stafford, F. E. (1964): J. Chem. Phys. **41**, 2207.

Skaarup, S., Boggs, J. E. and Skancke, P. N. (1974): Fifth Austin Symposium on Gas Phase Molecular Structure. Austin, Texas.

Skancke, P. N. and Boggs, J. E. (1973): Chem. Phys. Letters **21**, 316.

Smith, D., James, D. W. and Devlin, J. P. (1971): J. Chem. Phys. **54**, 4437.

Smith, D. W. and Hedberg, K. (1956): J. Chem. Phys. **25**, 1282.

Smith, H. W. and Lipscomb, W. N. (1965): J. Chem. Phys. **43**, 1060.

Smits, A. and Meijering, J. (1938): Z. phys. Chem. **B41**, 98.

Snelson, A. (1967): J. Chem. Phys. **46**, 3652.

Snelson, A. (1973): Infrared Studies of Vaporizing Molecules Trapped in Low Temperature Matrices, see Hallam (1973).

Snyder, R. G. and Hisatsune, I. C. (1957): J. Mol. Spectry. **1**, 139.

Solomonik, V. G., Girichev, G. V., Krasnov, K. S. and Zasorin, E. Z. (1973): All-Union Conference on Khimiya paroobraznykh neorganicheskikh soedinenii i protsessov paroobrazovaniya. Minsk.

Solomonik, V. G., Zasorin, E. Z., Girichev, G. V. and Krasnov, K. S. (1974): Izv. vysshikh uchebnykh zav. Khim. i Khim. Tekhnol. **17**, 136.

Sommer, A., White, D., Linevsky, M. J. and Mann, D. E. (1963): J. Chem. Phys. **38**, 87.

Spanbauer, R. N., Rao, K. N. and Jones, L. H. (1965): J. Mol. Spectry. **16**, 100.

Speiser, R. and Pierre, G. R. St. (1964): Fundamentals of Refractory Metal-Gaseous Environment Interaction. The Science and Technology of Tungsten, Tantalum, Molybdenum, Niobium and their Alloys (ed.: Promisel, N. E.). Pergamon, Oxford.

Spiridonov, V. P. (1974): Private communication.

Spiridonov, V. P., Akishin, P. A. and Tsirel'nikov, V. I. (1962): Zh. Strukt. Khim. **3**, 329.

Spiridonov, V. P., Brezgin, Yu. A. and Shakhparonov, M. I. (1971): Zh. Strukt. Khim. **12**, 1080.

Spiridonov, V. P., Brezgin, Yu. A. and Shakhparonov, M. I. (1972): Zh. Strukt. Khim. **13**, 320.

Spiridonov V. P. and Erokhin, E. V. (1969): Zh. Neorg. Khim. **14**, 636.

Spiridonov, V. P., Erokhin, E. V. and Brezgin, Yu. A. (1972): Zh. Strukt. Khim. **13**, 321.

Spiridonov, V. P., Erokhin, E. V. and Lutoshkin, B. I. (1971): Vest. Moskov. Univ. Ser. Khim. No. 3, 296.

Spiridonov, V. P., Khodchenkov, A. N. and Akishin, P. A. (1965a): Vest. Moskov. Univ. Ser. Khim. No. 6, 34.
Spiridonov, V. P., Khodchenkov, A. N. and Akishin, P. A. (1965b): Zh. Strukt. Khim. 6, 633.
Spiridonov, V. P., Khodchenkov, A. N. and Akishin, P. A. (1965c): Zh. Strukt. Khim. 6, 634.
Spiridonov, V. P. and Lutoshkin, B. I. (1970): Vest. Moskov. Univ. Ser. Khim. No. 6, 509.
Spiridonov, V. P. and Malkova, A. S. (1969): Zh. Strukt. Khim. 10, 332.
Spiridonov, V. P. and Mamaeva, G. I. (1969): Zh Strukt. Khim. 10, 133.
Spitsyn, V. I. and Shostak, V. I. (1949): Zh. Obshchei Khim. 9, 1801.
Srivastava, G. P. and Goyal, M. L. (1968a): J. Phys. B (Proc. Phys. Soc.) 1, 1312.
Srivastava, G. P. and Goyal, M. L. (1968b): Phys. Rev. 168, 104.
Srivastava, G. P. and Goyal, M. L. (1972): J. Phys. B. (Proc. Phys. Soc.) 4, 886.
Stalick, J. K. and Ibers, J. A. (1970): J. Am. Chem. Soc. 92, 5333.
Stephens, J. S. and Cruickshank, D. W. J. (1970a): Acta Cryst. B26, 222.
Stephens, J. S. and Cruickshank, D. W. J. (1970b): Acta Cryst. B26, 437.
Stevenson, D. P. and Schomaker, V. (1942): J. Am. Chem. Soc. 64, 2514.
Stork-Blaisse, B. A. and Romers, C. (1971): Acta Cryst. B27, 386.
Strieter, F. J., Templeton, D. H., Scheuerman, R. F. and Sass, R. L. (1962): Acta Cryst. 15, 1233.
Stucky, G. (1974): Stereochemical Properties of N-Chelated Alkali Metal Complexes. Polyamine-Chelated Alkali Metal Compounds. Advances in Chemistry Series, Vol. 130 (ed.: Langer, A. W.). American Chemical Society Publications, Washington, D. C.
Stull, D. R. (1966): JANAF Thermochemical Tables, Addendum. U. S. Department of Commerce, Washington, D. C.
Sugden, T. M. and Kenney, C. N. (1965): Microwave Spectroscopy of Gases. Van Nostrand, London.
Sundbom, M. (1966): Acta Chem. Scand. 20, 1608.
Sutter, D., Dreizler, H. and Rudolph, H. D. (1965): Z. Naturforsch. 20a, 1676.
Sutton, L. E. (1958): Tables of Interatomic Distances and Configuration in Molecules and Ions. Spec. Publ. No. 11, The Chem. Soc., London.
Sutton, L. E. (1965): Tables of Interatomic Distances and Configuration in Molecules and Ions, Spec. Publ. No. 18, The Chem. Soc., London.
Svec, H. J. and Junk, G. (1967): J. Am. Chem. Soc. 89, 2836.
Swanson, B. and Shriver, D. F. (1970): Inorg. Chem. 9, 1406.
Swanson, B., Shriver, D. F. and Ibers, J. A. (1969): Inorg. Chem. 8, 2182.
Szpiridonov, V. P. (1972): Kémiai Közlemények 37, 399.
Tanimoto, M., Kuchitsu, K. and Morino, Y. (1969): Bull. Chem. Soc. Japan 42, 2519.
Thompson, K. R. and Carlson, K. D. (1968): J. Chem. Phys. 49, 4379.
Tiemann, E., Hoeft, J., Lovas, F. J. and Johnson, D. R. (1974): J. Chem. Phys. 60, 5000.
Tiemann, E., Hoeft, J. and Törring, T. (1972): Fourth Austin Symposium on Gas Phase Molecular Structure. Austin, Texas.

Tiemann, E., Lovas, F. J. and Johnson, D. R. (1974): Fifth Austin Symposium on Gas Phase Molecular Structure. Austin, Texas.
Tkachev, V. V. and Atovmyan, L. O. (1972): Zh. Strukt. Khim. 13, 287.
Tolmachev, S. M. (1970): Dissertation, Moscow State University, Moscow.
Tolmachev, S. M. and Rambidi, N. G. (1971): Zh. Strukt. Khim. 12, 203.
Tolmachev, S. M. and Rambidi, N. G. (1972): Zh. Strukt. Khim. 13, 3.
Tolmachev, S. M., Zasorin, E. Z. and Rambidi, N. G. (1969): Zh. Strukt. Khim. 10, 541.
Trefonas, L. and Lipscomb, W. N. (1958): J. Chem. Phys. 28, 54.
Tremmel, J., Ivanov, A. A., Schultz, Gy., Hargittai, I., Cyvin, S. J. and Eriksson, A. (1973): Chem. Phys. Letters 23, 533.
Tubino, R. and Zerbi, G. (1969): J. Chem. Phys. 51, 4509.
Turley, J. W. and Rinn, H. W. (1969): Inorg. Chem. 8, 18.
Tuseev, N. I., Zasorin, E. Z. and Spiridonov, V. P. (1974): Fifth All-Union Conference on Fizicheskie i matematicheskie metody v koordinatsionnoi khimii. Kishinev.
Tyler, J. K., Cox, A. P. and Sheridan, J. (1959): Nature 183, 1182.
Ueno, K. (1941): J. Chem. Soc. Japan 62, 990.
Ugarov, V. V., Ezhov, Yu. S. and Rambidi, N. G. (1973a): Zh. Strukt. Khim. 14, 359.
Ugarov, V. V., Ezhov, Yu. S. and Rambidi, N. G. (1973b): Zh. Strukt. Khim. 14, 548.
Ugarov, V. V., Tolmatchev, S. M., Ezhov, Yu. S. and Rambidi, N. G. (1972): Fourth Austin Symposium on Gas Phase Molecular Structure. Austin, Texas.
Ugarov, V. V., Vinogradov, V. S., Zasorin, E. Z. and Rambidi, N. G. (1971): Zh. Strukt. Khim. 12, 315.
Ugi, I. and Ramirez, F. (1972): Chemistry in Britain 8, 198.
Ukshe E. A. (ed.) (1966): Stroenie rasplavlennykh solei. Mir, Moscow.
Vandoorne, W., Cordes, A. W. and Hunt, G. W. (1973): Inorg. Chem. 12, 1686.
Veniaminov, N. N., Ustynyuk, Yu. A., Alekseev, N. V., Ronova, I. A. and Struchkov, Yu. T. (1970): J. Organometal. Chem. 22, 551.
Veniaminov, N. N., Ustynyuk, Yu. A., Alekseev, N. V., Ronova, I. A. and Struchkov, Yu. T. (1971a): Dokl. Akad. Nauk S. S. S. R. 199, 346.
Veniaminov, N. N., Ustynyuk, Yu. A., Alekseev, N. V., Ronova, I. A. and Struchkov, Yu. T. (1971b): Zh. Strukt. Khim. 12, 952.
Veniaminov, N. N., Ustynyuk, Yu. A., Alekseev, N. V., Ronova, I. A. and Struchkov, Yu. T. (1972): Zh. Strukt. Khim. 13, 136.
Veniaminov, N. N., Ustynyuk, Yu. A., Struchkov, Yu. T., Alekseev, N. V. and Ronova, I. A. (1970): Zh. Strukt. Khim. 11, 127.
Vilkov, L. V., Khaikin, L. S., Zhigach, A. F. and Siryatskaya, V. N. (1968): Zh. Strukt. Khim. 9, 889.
Vilkov, L. V., Mastryukov, V. S. and Akishin, P. A. (1963): Zh. Strukt. Khim. 4, 323.
Vilkov, L. V., Mastryukov, V. S., Akishin, P. A. and Zhigach, A. F. (1965): Zh. Strukt. Khim. 6, 447.

Vilkov, L. V., Mastryukov, V. S., Zhigach, A. F. and Siryatskaya, V. N. (1966): Zh. Strukt. Khim. **7,** 883.

Vilkov, L. V., Mastryukov, V. S., Zhigach, A. F. and Siryatskaya, V. N. (1967): Zh. Strukt. Khim. **8,** 3.

Vilkov, L. V., Rambidi, N. G. and Spiridonov, V. P. (1967): Zh. Strukt. Khim. **8,** 786.

Vinogradov, S. N. and Linnel, R. H. (1971): Hydrogen Bonding. Van Nostrand Reinhold Co., New York.

Vladimiroff, T. (1972): J. Am. Chem. Soc. **94,** 8250.

Vranka, R. G. and Amma, E. L. (1967): J. Am. Chem. Soc. **89,** 3121.

Vrieland, G. E. and Stull, D. R. (1967): J. Chem. Eng. Data **12,** 532.

Wade, K. (1968): Chemistry in Britain **4,** 503.

Wade, K. (1972): J. Chem. Educ. **49,** 502.

Wallwork, S. C. (1959): Proc. Chem. Soc., 311.

Wang, H. K. (1965): Acta Chem. Scand. **19,** 879.

Webb, D. V. and Rao, K. N. (1968): J. Mol. Spectry. **28,** 121.

Webster, M. and Keats, S. (1971): J. Chem. Soc. A., 836.

Weidlein, J. (1969): J. Organometal. Chem. **17,** 213.

Weidlein, J. and Krieg, V. (1968): J. Organometal. Chem. **11,** 9.

Weltner, W., Jr. and McLeod, D., Jr. (1965): J. Mol. Spectry. **17,** 276.

Westrick, R. and MacGillavry, C. H. (1941): Rev. Trav. Chim. **60,** 794.

Wharton, L., Berg, R. A. and Klemperer, W. (1963): J. Chem. Phys. **39,** 2023.

Wheatley, P. J. (1959): The Determination of Molecular Structure. Clarendon Press, Oxford.

White, D., Mann, D. E., Walsh, P. N. and Sommer, A. (1960): J. Chem. Phys. **32,** 488.

White, D., Walsh, P. N. and Mann, D. E. (1958): J. Chem. Phys. **28,** 508.

Wilkinson, G. and Cotton, F. A. (1959): Progress in Inorganic Chemistry. **1.** Interscience Publishers, New York.

Wilkinson, G., Rosenblum, M., Whiting, M. C. and Woodward, R. B. (1952): J. Am. Chem. Soc. **74,** 2125.

Williams, J. E. and Murrell, J. N. (1971): J. Am. Chem. Soc. **93,** 7149.

Williams, R. E. (1970): Carboranes. Progress in Boron Chemistry. Vol. 2 (ed.: Brotherton, R. J. and Steinberg, H.). Pergamon Press, Oxford.

Wilson, E. B., Jr. and Lide, D. R., Jr. (1955): Microwave Spectroscopy. Determination of Organic Structures by Physical Methods (ed.: Braude, E. A. and Nachod, F. C.). Academic Press, New York.

Winnewisser, G. (1972): J. Mol. Spectry. **41,** 534.

Wollrab, J. E. (1967): Rotational Spectra and Molecular Structure. Academic Press, New York.

Wollrab, J. E. and Laurie, V. T. (1969): J. Chem. Phys. **51,** 1580.

Wong, C. and Schomaker, V. (1957): J. Phys. Chem. **61,** 358.

Wong, C.-H., Lee, T.-Y., Chao, K.-J. and Lee, S. (1972): Acta Cryst. **B28,** 1662.

Zachariasen, W. H. and Plettinger, H. A. (1961): Acta Cryst. **14,** 229.

Zakharkin, L. I., Stanko, V. I. and Brattsev, V. A. (1964): Dokl. Akad. Nauk S. S. S. R. **155,** 1119.

Zalkin, A. and Templeton, D. H. (1964): J. Chem. Phys. **40,** 501.

Zasorin, E. Z. (1965): Dissertation, Moscow State University, Moscow.
Zasorin, E. Z., Ivanov, A. A., Spiridonov, V. P., Hargittai, I. and Hargittai, M. (1973): All-Union Conference on Khimiya paroobraznykh neorganicheskikh soedinenii i protsessov paraoobrazovaniya. Minsk.
Zasorin, E. Z. and Rambidi, N. G. (1967a): Zh. Strukt. Khim. **8,** 391.
Zasorin, E. Z. and Rambidi, N. G. (1967b): Zh. Strukt. Khim. **8,** 591.
Zasorin, E. Z., Rambidi, N. G. and Akishin, P. A. (1963): Zh. Strukt. Khim. **4,** 910.
Zmbov, K. F., Hastie, J. W., Hauge, R. and Margrave, J. L. (1968): Inorg. Chem. **7,** 608.
Zvonkova, Z. V. (1956): Kristallografija **1,** 73.